AI for Business

No-Hype Guide to Apply Artificial Intelligence at Work: Navigate Trends, Avoid Bias, & Lead with Confidence to Future-Proof Yourself & Your Team with New Skills & Wins

George Munson

GL Digital Publishing LLC

AI FOR BUSINESS

First Edition

Contents

Introduction

Let's be real: Artificial intelligence is everywhere, on your news feed, in your group chats, and all over the office. But every time you try to learn about it, you're hit with technical jargon, wild predictions, or endless "revolutionary" headlines. It's overwhelming. It's confusing. And AI is either going to fix everything or ruin everything, sometimes both.

I get it. I've been there, too, staring at articles that use words like "deep learning" and "neural networks" as if everyone should already know what they mean. I've watched people get excited about AI tools at work, only to get stuck after the first login screen. I've also seen the worry, that nagging fear of being left behind while everyone else seems to "get" AI.

That's why I wrote this book. My goal is straightforward: to help you cut through the noise and effectively use AI in ways that benefit you, both at work and at home. Not just to feel smarter but to get better results, save time, and feel more confident in the choices you make. I'm passionate about making AI accessible, not just something you hear about.

Let's talk about those everyday frustrations. You may have tried a new AI-powered tool and felt lost after the first few steps. Perhaps you've wondered if you're "doing it wrong" because everything seems more complicated than promised. You may want someone to explain things in plain English without making you feel like you need a PhD to keep up. If any of this sounds familiar, you're in the right place.

This book is not just another guide telling you AI will change your life, then leaving you to figure out the details. I've packed it with step-by-step projects, real-world examples, and quick wins you can use right now. You'll see how AI can help you write better emails, sort through information faster, make smarter decisions, and even find new career opportunities. Every chapter is built around practical outcomes, not hype, empowering you to take control of your AI journey.

So, what makes this book different? I believe in showing, not just telling. You'll find visual guides, checklists, and "try it now" sections, all designed for busy individuals. No fluff, no filler, just what you need to start, experiment, and see results. I've also ensured that you don't need a technical background. Zero prerequisites. If you've ever used a smartphone, you're ready. This book is designed with you in mind, ensuring you feel included and considered.

Let me give you a sneak peek at the "AI for Real Life" Blueprint we'll use together:

- **Foundations in Plain English:** You'll understand what AI is (and isn't) without needing to decode technical language.

- **Guided Projects:** Step-by-step walkthroughs for using AI in your daily work and personal life.

- **Industry Playbooks:** Real use cases from different fields, so you can see how others are already winning with AI.

- **Ethics in Action:** Simple frameworks so you can use AI responsibly and spot bias before it becomes a problem.

- **Upskilling Paths:** Clear roadmaps for growing your AI skills, whatever your starting point.

- **Trends That Matter:** The latest updates without the hype, so you know what's worth your time.

Are you worried about getting lost or stuck? Don't be. I've built this book for real people with real questions, not tech insiders. Each chapter offers instant, shareable wins, little things you can try right away to build your confidence. And if you're worried about feeling alone, you're not. You'll have access to an online community where you can ask questions, share your wins, and learn from others on the same path.

Maybe you're thinking, "What if I just don't get it?" or "What if I mess up?" That's okay, too. This book is designed with beginners in mind. You're allowed to be curious, cautious, even skeptical. I'll never talk down to you or expect you to be an expert before you start. Your questions and doubts are welcome here.

Here's what you can expect as you read: You'll walk away with tools you can use on Monday morning, not just theories or buzzwords. You'll receive communication tips for discussing AI with non-technical friends and coworkers. You'll see how to spot trends that matter for your career and how to avoid the ones that waste your time. Most of all, you'll gain confidence that you can use AI at your own pace, in your way.

You're not doing this alone. If you take away one thing from this introduction, let it be this: AI isn't just for techies, CEOs, or "geniuses." It's for you. It's a tool, not a threat. With the right approach, you can utilize it to solve real problems, make better decisions, and open new doors in both your professional and personal life.

So, let's get started. No hype. No jargon. Just fundamental skills, real stories, and real results together.

Chapter One

Demystifying AI From Buzzword to an Everyday Tool

Y ou've probably seen headlines warning of AI "taking over the world," only to get a notification from your local pizza place's chatbot minutes later. The gap between AI hype and its real daily impact is vast. Maybe you've heard coworkers talk about "machine minds" or seen jokes about robots writing novels. Still, your voice assistant can't even find the nearest ATM. The noise can be overwhelming. If you're skeptical or annoyed by all the talk, you're not alone.

I wrote this chapter because even well-informed adults are often stuck between thinking AI is either pure science fiction or a looming threat out to steal jobs, privacy, or peace of mind. My goal is to cut through the drama and explain what AI is, what it can and can't do, and how it's already part of your life. You don't need to understand tech jargon. If you use email, shop online, or get phone directions, AI has already affected you, sometimes so subtly you barely notice.

What AI Is and Isn't

Here's the truth: Artificial intelligence isn't magic. There are no sentient robots or machines with genuine feelings. At its core, AI utilizes mathematical and computational methods to enable machines to perform tasks that we associate with human intelligence, such as recognizing speech, translating text, sorting photos, predicting movie choices, or suggesting meeting times. Think of AI as a specialized tool that identifies patterns, calculates quickly, and predicts outcomes based on data.

Take *Siri* or *Alexa* as examples. They're not real people and don't truly "understand" you. They match your words to probable commands using a complex pattern recognition. When you say "Call Mom," there's no sentiment involved; it's just algorithms mapping your words to an action. Email spam filters work similarly. They aren't "reading" for meaning but use models to spot likely spam based on words, phrases, and past behaviors.

A central myth is that AI is on the verge of becoming super intelligent or conscious. It's a favorite movie plot, but not reality. Most real-world AI is "narrow": it excels at one task exceptionally well, such as identifying faces or recommending playlists, but struggles to operate outside that specific task. Even advanced chatbots won't start writing poetry out of genuine inspiration or decide to become chefs. They stay within strict limits, performing only what they're trained to do.

You'll see plenty of gadgets marketed as learning your every need or thinking for you. In reality, many consumer AIs are overhyped. "Smart" fridges often struggle to manage your groceries reliably, and fitness trackers may misinterpret your health due to a missed workout. It's healthy to be skeptical. If an ad or product promises AI can do everything but tuck you in at night, take it with a grain of salt.

What AI can't do is just as important. Machines don't understand emotions or context. Suppose you vent to a chatbot about a tough day. In that case, it responds with standard lines; it can't feel empathy or get nuances. Its replies are based on patterns in massive datasets, not a proper understanding. Understanding these limitations is crucial in forming realistic expectations about AI. By being aware of these limitations, you can gain a deeper understanding of the capabilities and potential of AI.

This gap is evident when using customer service bots, as they may provide generic answers that appear correct but fail to address your actual question because they recognize keywords rather than your intent or emotional state. There is a significant difference between sounding human and truly being human, and current AI hasn't bridged that divide.

In daily life, AI often operates behind the scenes. Is *Netflix* recommending your next show? That's AI analyzing your viewing history and comparing it to millions of others to spot trends. Instagram's excellently targeted ads? Algorithms track your likes, follows, and even how long you pause on a post. Map apps offering tailored restaurant lists? That's AI making predictions based on your past habits and similar users. Even your email spam filter uses AI to filter out junk by scanning for risks in incoming emails.

In all these cases, AI isn't thinking by itself; it's applying learned rules from vast amounts of data to make educated guesses.

Take a Moment for a Quick Reflection

Pause to note three ways AI has appeared in your day so far, in a *Spotify* playlist suggestion, a targeted ad, or email autocomplete. Spotting these everyday uses demonstrates how AI impacts daily routines in practical terms.

Ultimately, artificial intelligence is less about creating digital minds and more about building tools that spot patterns and make predictions to help us solve problems faster and more efficiently. It's not about sentience; it's about practical benefits. Recognizing this helps set realistic expectations and distinguish between hype and genuine innovation. AI is here to make our lives easier, not to replace us. Whether it's a *Spotify* playlist suggestion, a targeted ad, or email autocomplete, these everyday uses of AI demonstrate how it impacts our daily routines in practical and beneficial ways.

Decoding the Tech Lingo of AI vs. Automation vs. Machine Learning

Tech conversations can become blurry quickly, especially when people use terms like "AI," "automation," and "machine learning" interchangeably. This happens in offices, boardrooms, and even over coffee. However, these terms are not interchangeable, and mixing them up can lead to headaches, missed opportunities, and sometimes expensive mistakes. Here's the simple breakdown: Automation is about following instructions without thinking; machine learning is about identifying patterns and making predictions based on data; AI is the broad umbrella that encompasses both, as well as more advanced abilities, such as understanding language or recognizing images. Automated systems do what they're told every time, with no surprises. Picture a robotic arm on a car assembly line, tightening bolts the same way all day long. That's Automation, reliable, tireless, but not adaptable. Now, think about machine learning. This is a branch of AI where computers actually "learn" from data. For example, *Netflix* uses machine learning to predict what you'll want to watch next based on your viewing history and the preferences of others. The more you watch, the better its guesses get. It doesn't just follow static instructions; it adapts as new data comes in.

Artificial intelligence, on the other hand, is a broader field. It encompasses any technology that attempts to mimic human intelligence, even in a limited manner. A virtual customer service agent that chats with customers, answers questions, and may transfer you to a real person if it gets stumped — that's AI in action. It can understand requests, remember past interactions, and make decisions on the fly. Not all AI uses machine learning, and not all Automation is intelligent enough to qualify as AI.

Why do these distinctions matter? The mix-ups aren't just academic. They play out in the workplace whenever someone evaluates new software or pitches an upgrade to management. Imagine your company is looking to speed up invoice processing. One team member suggests a rule-based workflow automation that flags invoices exceeding a certain amount for review, offering a simple, predictable, and low-cost solution. Another proposes a machine learning-driven analytics tool that can identify unusual spending patterns or even detect fraud by learning from years of historical data. If no one knows the difference, you risk buying something that doesn't solve your actual problem or paying for complexity you don't need.

To help with this, here's a scenario I've seen more than once: A business leader wants to "add AI" to their sales process. They hear about chatbots and imagine an intelligent assistant handling all customer queries. The IT team rolls out a rule-based chatbot that can only answer set questions. Customers quickly become frustrated when they request anything unexpected. The leader expected an adaptive AI agent but got basic Automation instead. That's why knowing what each technology offers (and what it can't do) is crucial for setting expectations and making wise investments.

How to Decode the Differences of Automation vs. AI

- **Automation** →
  - ○ Follows set rules for →
    - robotic arms
    - payroll deposit scripts
    - invoices and billing
- **Artificial Intelligence** →
  - ○ Mimics limited human abilities →
    - virtual agents like *Google Assistant*
    - intelligent assistants like *Amazon Alexa*
  - ○ Includes machine learning →
    - spam filtering of email
    - facial recognition
  - ○ Learns from data →
    - *Netflix* suggestions
    - credit risk scoring

Case studies make this real. In HR, automated payroll processing systems eliminate repetitive monthly tasks by following strict rules. No intelligence is needed, just speed and accuracy. In marketing, AI-powered sentiment analysis tools scan social media posts to gauge the public's mood about a brand, using language understanding and, in some cases, tone detection, both AI techniques. Banking provides another window: Modern credit scoring often uses machine learning to sift through thousands of variables, including income history, spending patterns, and even the time of day transactions occur, to predict who is likely to repay a loan. These models improve as they process more data, becoming sharper at spotting risk factors that traditional Automation would miss.

Compare two customer support solutions: A rule-based chatbot operates based on a decision tree, "If a user says X, respond with Y." It gets stuck if someone asks a question outside its script. A conversational AI assistant goes further; it can parse language more flexibly, remember context from earlier in the chat, and even escalate tricky issues to humans when necessary. That's where AI moves beyond rigid Automation.

Still, it's easy to blur the lines, especially since many tools now combine these technologies for optimal results. For instance, an automated payroll system might quietly use machine learning to flag suspicious entries for review, adding a layer of intelligence to routine Automation.

When you break it down, Automation is the tireless worker who never improvises; machine learning is the analyst who notices patterns and adapts; AI is the all-rounder who sometimes chats with customers, sometimes analyzes data, and sometimes follows instructions, depending on what's needed. Knowing which tool does what helps you choose wisely and avoid falling for the latest buzzword when making decisions at work or on your projects.

The Real Story Behind Algorithms and Data Reveals How AI "Learns"

Peeling back the curtain on how AI "learns" can help you view it as something practical rather than magical. Imagine teaching a child to distinguish between cats and dogs in photos. You'd begin with a stack of snapshots, some with cats, others with dogs. For each picture, you'd say, "This one's a cat," or "That's a dog." The child starts to look for clues: pointy ears, whiskers, floppy tails, furry faces. Over time, as you show more examples and offer feedback, the child does less guessing and more recognizing. This is how most AI systems get trained, through repetition, exposure, and correction.

AI eats data for breakfast. The process starts by feeding it thousands, or even millions, of labeled examples. In our animal scenario, each image is accompanied by a tag: "cat" or "dog." The AI scans these images, searching for patterns. Perhaps cats have smaller noses or sit in different positions. It uses these subtle distinctions to build rules for future predictions. These rules aren't written by hand; instead, the AI generates them autonomously as it continues to process data. The more quality examples it gets, the more accurate (and confident) its predictions become.

Here's where the role of data quality, quantity, and diversity steps in. Suppose you only ever show the AI pictures of tabby cats and golden retrievers. In that case, it might struggle to recognize a Sphynx cat or a dachshund. If your photos all have the same background or lighting, real-world photos will trip it up. When companies use AI for hiring and only train models on outdated resumes from a single demographic, the tool may inadvertently pick up unintentional biases, skewing toward certain schools or experiences while overlooking equally qualified candidates from other backgrounds. Incomplete or skewed data sets create blind spots. When an AI is not exposed to sufficient real-world data during training, it tends to make mistakes when faced with new situations.

Imagine a company using AI to predict which products will sell best next month. Suppose they only feed it sales data from the holiday seasons. In that case, the AI may predict December-style demand in July, a classic case of missing context due to incomplete data. The lesson is clear:

Garbage in → garbage out → a pile of trash

Great data makes → great AI → solid results

Poor data leads to → unreliable results → confusion & error

Now, regarding algorithms and models, they may sound intimidating, but they're not. Think of an algorithm as a recipe, a set of rules for processing information. A model is like a finished cake baked from that recipe using real ingredients (data). When you send in new photos, the trained model uses what it's learned to decide: cat or dog? However, models don't emerge ideally on day one. They're tested and tweaked. AI engineers divide the data into training sets (for training) and test sets (for evaluating progress). If the model mislabels a French Bulldog as a cat, it adjusts the parameters and tries again. Over time, this back-and-forth process sharpens performance.

Spam filters are a classic example. Engineers review thousands of emails, marking them as either "spam" or "not spam." The filter learns word patterns, sender info, and even subject line quirks that often signal junk mail. When you get a new email, the spam filter predicts which pile it belongs in by comparing it to those learned patterns.

How does this learning change what you experience? There are several approaches, and each shape yields different results. With supervised learning (like our animal example), humans provide both the questions and answers: Here's the photo; here's what it shows. The AI learns directly from labeled data. Unsupervised learning is like giving someone a box of puzzle pieces without showing them the final picture. The system must group similar items by clustering customer types based on spending habits, without requiring explicit answers upfront.

Then there's reinforcement learning: Think of a dog learning tricks with treats. The AI attempts actions (such as moving left or right in a game), receives rewards for effective moves, and learns through trial and error which strategies are most effective. This approach powers technologies like self-driving cars, which need to adapt quickly to unexpected events on the road.

Recommendation engines, such as those used by *Spotify* playlists or *YouTube* video suggestions, improve as they collect more data from users' likes and skips. Each interaction helps fine-tune future predictions, allowing you to see content that better suits you over time. Self-driving vehicles process millions of driving scenarios to learn how to handle everything from rain-slick streets to four-way stops; their algorithms adapt as they encounter new road situations.

Customer service chatbots also "learn" after launch. When users ask questions that stump the bot, developers review those interactions and update its knowledge base so that it responds more accurately next time. Each real-world use helps refine the model.

AI learning isn't magic; it's iteration, feedback, and massive data crunching done at machine speed. When it works well, it feels seamless; when it fails, it's usually due to insufficient data or unrealistic expectations about what the system was ever intended to do. Understanding this process gives you an eye for what's possible and where things might go off track; no technical degree is required.

A Beginner's Guide with Diagrams for Visualizing Neural Networks

Picture a neural network as an assembly line in a factory. Raw materials enter one end and are transformed into finished products through a series of stations. Similarly, a neural network processes data, like images, audio, or spreadsheet numbers, through layers: input, hidden, and output. Each layer acts as a checkpoint, gradually refining the data to generate a final answer or prediction. For example, when recognizing handwritten numbers, the input layer receives pixel data from a scanned image, passing the brightness of each pixel as a number to the next stage.

The transformation happens in the hidden layers between input and output. Think of these as specialized workers: instead of physical parts, they're crunching numbers and hunting for patterns to answer, "What number is this?" Hidden layers use formulas and weights to refine the data, much like adjusting dials based on previous learning. More hidden layers mean deeper, more refined pattern recognition.

At the end sits the output layer, where the system produces its final result. In the handwritten digit example, the network picks from ten options (0–9); in speech-to-text, it outputs recognized words. Data only ever moves forward, from one layer to the next, so this approach is called "feedforward," like products moving down a conveyor belt.

Let's get concrete using the handwritten digits example. When you take a photo of a scribbled "3," each pixel value goes to the input layer. The first hidden layer might look for basic shapes, such as lines, curves, and angles. Later layers combine these elements into larger features, like loops or intersections. By the time data hits the output layer, the network has enough clues to call your scribble a "3 confidently." This process happens in milliseconds.

Neural networks aren't just academic experiments; they power everyday technology. Face unlock on your phone uses a neural network that analyzes thousands of facial points, handling changes in lighting or expression through its hidden layers. Likewise, voice assistants use neural networks to convert speech to text by processing audio through multiple layers to accurately predict the correct words.

However, neural networks have challenges. Their power comes from sifting through vast amounts of messy data to identify patterns that people might miss, making them ideal for facial recognition or speech-to-text applications. This is where traditional programming struggles to excel. But neural networks require massive amounts of data to learn correctly. Training on only a few examples means the network won't grasp key differences, such as identifying every dog breed from just five photos.

Another notable downside is their opacity. Often referred to as "black boxes," neural networks can be challenging to understand and interpret. While you can see the input and the output, understanding the internal reasoning is hard, even for experts. If your phone refuses to unlock one day or your command is misheard, it's rarely obvious why. This lack of transparency can frustrate users and complicate development and troubleshooting.

Take the portrait mode on your smartphone, for example. The neural network identifies what in your photo is a person and what is the background. When it works, faces are sharp, and backgrounds blur nicely; when it fails, ears might blur with the scenery, or edges can look unnatural. These blemishes reveal both the power and limitations of neural networks. They rely entirely on the quality and quantity of their training data.

Imagine that assembly line again, but now each worker is blindfolded, except for a peephole that shows just their task. The finished product depends on every step working correctly, but seeing exactly what happened at each station is nearly impossible from the outside. That's the mystery and strength of neural networks: powerful pattern detectors with some hidden inner workings.

Simple Neural Network

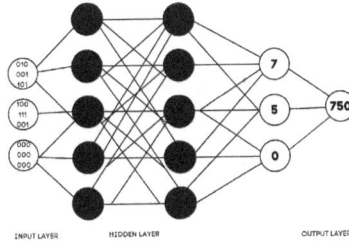

In real life, neural networks enable features such as phone unlocking, photo auto-tagging, translation apps, and even medical scan analysis. From snapping selfies to dictating messages, neural networks work silently and swiftly, improving as they encounter more data. Their strength lies in processing complexity at scale, while their main weaknesses are when available data is limited or when humans require clear explanations for each decision.

From Black Box to Glass Box to Make AI Explainable for Everyone

AI does some of its most impressive work behind the scenes, but that's also where things can get a little murky. You hear people call advanced AI models "black boxes" for a reason; they take in your data, process it in complex, layered steps, and spit out an answer, but you rarely see how they got there. It's as if you hand over your question to a magician, and a solution pops out with no explanation. This isn't just frustrating; it creates real problems for trust and accountability. Imagine a hospital relying on AI to suggest a diagnosis. Still, the doctor can't figure out why the system flagged one patient as high risk and another as low. That lack of transparency can lead to confusion or even harm. If you're being denied a mortgage or your medical exam results spark a sudden change in treatment, you want to know why. Without clear explanations, people question whether the technology is fair or even safe.

That's where explainable AI, or XAI, comes in. Instead of keeping decisions locked away, XAI aims to pull back the curtain and show how and why an AI system makes a choice. Think of it as swapping out tinted glass for something transparent. One popular XAI tool is a feature importance chart, essentially a ranked list that shows which factors have the most significant impact on a decision. Say you apply for a loan and get rejected. Instead of an unhelpful "You don't qualify," an XAI-powered system might break down the verdict: "Your income level was the biggest factor (35%), followed by your credit history (25%) and your debt-to-income ratio (20%)." Suddenly, the process feels less like a lottery and more like something you can understand and improve upon next time.

These explanations aren't just nice to have; they're becoming critical in business and society. Companies face tough questions about bias and discrimination, particularly when utilizing AI for hiring or lending purposes. Suppose an algorithm identifies patterns that inadvertently disadvantage certain groups. In such cases, hidden logic can perpetuate unfairness without anyone realizing it until the damage is already done. Regulators are starting to step in. For example, the *General Data Protection Regulation* (GDPR) in Europe requires organizations to explain decisions made by automated systems, particularly when those decisions significantly impact individuals' lives. Suppose a bank uses AI to approve or deny credit. In that case, it must provide a meaningful explanation rather than hiding behind "the algorithm decided." Transparent logic helps identify unfair patterns early, enables audits, and builds user confidence that systems are not making arbitrary or biased choices (see *Explainable AI (XAI): The Complete Guide* (2025) - Viso Suite).

Ethics and business outcomes are deeply connected here. Imagine a hiring platform that automates the screening of resumes. Suppose HR managers can see exactly which skills or keywords tipped the decision for each candidate. In that case, they're more likely to trust both the process and the results. Additionally, candidates receive feedback they can use and no longer feel like their application has fallen into a black hole. When people understand why things happen, they feel respected, whether they're customers, employees, or partners.

For those of us who aren't building these tools from scratch, demanding explainability is still entirely within reach. You don't need to learn code or become a data scientist to push for clear answers. Any time you're considering a new AI-powered product, whether it's a vendor pitching recruiting software or an internal team rolling out analytics, come prepared with questions that cut through the fog. Here's a practical list to keep handy:

- Can you show me which factors weighed most heavily in this decision?

- If the system makes a mistake, how does it correct itself?

- How do you check for bias or unintended side effects?

- Can I see examples of how explanations are delivered to end users?

- What level of transparency do I get compared to regulators or auditors?

- If I don't understand the explanation, who helps me make sense of it?

Use what I call the "Explain it to me like I'm five" test. If the answer sounds like more jargon or just circles back to "the AI knows best," push for something clearer. You should never feel bad about asking for plain English, especially when technology impacts your life or work.

Transparency isn't just about compliance; it's about building trust, so you want to use these tools and feel confident recommending them to others. When teams ask thoughtful questions up front, they avoid surprises down the road. If everyone involved knows what's happening under the hood, even at a high level, it's easier to spot problems early and make better decisions together.

The future of AI isn't just about smarter predictions or faster answers. It's about ensuring that those answers are fair, understandable, and open to scrutiny. When we transition from black-box thinking to glass-box clarity, everyone benefits, users gain control, businesses foster loyalty, and technology becomes more responsible with every step forward.

Chapter Two

AI in Action, Real-World Case Studies Across Industries

How AI Is Transforming Healthcare From Diagnosis to Drug Discovery

Suppose you're waiting at your doctor's office, anxious about a persistent cough. Instead of enduring rounds of inconclusive tests, your doctor uploads your CT scan to a secure AI system. Within seconds, the AI, trained on millions of similar cases, flags a tiny spot that a human might miss, prompting a closer look. This isn't futuristic fiction; hospitals worldwide now utilize AI to analyze complex radiology images, enabling specialists to identify tumors, fractures, or infections more quickly and accurately. The benefit isn't just technological flash, it's about saving crucial minutes in life-or-death situations and reducing errors that could haunt patients and families. This AI doesn't just analyze data; it provides a sense of relief and reassurance, knowing that every possible detail has been considered.

AI's strength in pattern recognition is also transforming disease prediction. Software now combs through years of electronic health records, flagging early warning signs for illnesses like diabetes or heart disease, often before obvious red flags appear. These predictive tools analyze dozens of variables, including age, lifestyle, lab results, and even location, to promptly alert care teams and prevent crises. This proactive approach leads to fewer hospital stays, lower costs, and a better chance of intercepting chronic disease before it progresses.

However, AI is revolutionizing more than just diagnosis; it's dramatically accelerating drug discovery. Traditional methods require screening thousands of compounds in the lab, hoping to find one that works. Machine learning models now simulate how molecules will interact with the human body, saving time and expense. A landmark example is DeepMind's AlphaFold, which cracked the challenge of predicting protein 3D structures. This mystery stymied scientists for decades. With this knowledge, researchers can design drugs with greater precision, targeting diseases ranging from rare disorders to cancer by matching drugs directly to the relevant proteins. Virtual screening platforms can scan millions of drug candidates overnight, identifying the most promising ones for further testing, especially critical for rare diseases where time is precious and cases are few.

Healthcare is also about logistics and supporting care teams. Anyone who's waited for hours for an appointment or seen nurses buried in paperwork knows the system's inefficiencies. AI is being deployed to automate surgery scheduling, aligning staff and operating rooms to urgent needs in real-time, which reduces waitlists and helps balance workloads to prevent burnout. Automated patient triage chatbots now handle the initial round of symptom checks, routing patients to appropriate care and freeing doctors to focus on complex cases.

Of course, medical AI faces unique challenges. Healthcare tools must clear stringent regulatory hurdles before they can be used. In the US, the Food and Drug Administration (FDA) reviews AI diagnostics just as rigorously as it reviews new drugs, demanding evidence that they improve care without introducing additional risks. Transparency is crucial as doctors and patients must understand how an algorithm reaches its conclusions, especially if it affects treatment decisions or delivers significant news. Bias is another concern: if trained primarily on data from one demographic, an AI might overlook significant signs in patients from other backgrounds. Regulators now require proof that AI tools are effective across diverse populations.

Consider...

Consider your most recent healthcare experience, such as a clinic visit or routine blood test. List three ways smarter scheduling, quicker test results, or more tailored risk assessments could have made the process smoother or more accurate for you or a loved one. Suppose you work in healthcare or assist someone who does, identify routine frustrations that AI-powered solutions could help resolve. In that case, even minor improvements accumulate when multiplied across many patients each month.

Rather than replacing healthcare professionals, AI equips them with sharper tools, allowing clinicians to focus on what truly matters: human connection, empathy, and lifesaving care. This empowerment inspires them to push the boundaries of what's possible in healthcare, knowing that AI is there to support and enhance their efforts.

AI in Marketing for Personalization Without Creepiness

Do you know that feeling when an online store seems to know what you want before you do? That's not magic, it's AI working behind the scenes, quietly segmenting you into groups based on your clicks, likes, and even the time you typically shop. AI-powered marketing teams no longer blast one-size-fits-all ads. Instead, they use algorithms to predict what catches your interest, adjusting product recommendations in real-time. When you shop for sneakers, and suddenly every ad is about running gear, that's a machine learning model at work. The trick is making these suggestions feel helpful, not invasive. Marketers tread a thin line between "Wow, that's useful" and "How did they know that?" By analyzing patterns, such as the pages you visit, what you leave in your cart, and even how quickly you scroll, AI personalizes content; however, the best brands keep it subtle. You might notice product suggestions that suit your taste or emails with subject lines that arrive at precisely the right time for you to open them. This level of convenience and efficiency leaves you feeling satisfied and content with your online experience.

AI also supercharges content creation and testing. Gone are the days of guessing which headline will convert best. Now, marketers feed dozens of options into AI copy generators and instantly get variants tested against live audiences. Imagine A/B testing a headline: AI tracks which version gets more clicks and automatically shifts traffic to the winner. That's not all, automation tools now select which images to display on social media posts based on what's trending among your demographic. Say you run a local coffee shop; an AI tool might recommend posting cozy autumn shots to boost engagement as the weather cools in your area. These systems adapt constantly, learning what works and discarding what doesn't, so campaigns get smarter over time without endless manual tweaks.

Mapping the customer journey has always been challenging, as people switch devices, abandon carts, and move between social media and websites. AI changes the game by stitching those touchpoints together. Predictive analytics allow brands to forecast when you might cancel a subscription or lose interest in a service. For example, streaming platforms use churn prediction models that flag users at risk of leaving based on skipped episodes or search habits. Next-best-offer engines in retail go a step further: they suggest just the right discount or product bundle, often before you realize you want it. Sometimes, these nudges are so well-timed that they feel almost psychic. Still, it's just fast data crunching and pattern recognition working at scale.

Of course, there's a flip side. Personalization can get creepy fast if it's not handled with care. Responsible marketers pay close attention to ethics and privacy. The most innovative companies prioritize transparency by clearly disclosing how your data is used and providing you with genuine choices about what is tracked. In Europe, for example, GDPR-compliant personalization means that every targeted campaign begins with explicit consent from the individual. You'll see clear options to opt in or out of AI-driven recommendations, and brands must explain their data practices in language anyone can understand (see "AI and GDPR compliance in marketing practices," 2023). This isn't just a legal box to check; it's about trust. When people know they can control how brands use their data, and that algorithms aren't making wild leaps based on private info, they're more likely to welcome personalization.

Quick Checklist for Spotting Ethical AI Marketing in Action

- Do you see clear opt-in or opt-out choices for personalized content?

- Can you easily find out why a brand is recommending something to you?

- Does the company explain how your data is used in plain language?

- Are recommendations genuinely helpful, or do they feel intrusive?

- Is there an easy way to adjust or delete your preferences?

If more brands followed these simple rules, personalization would feel less like digital stalking and more like genuine service. AI gives marketers incredible power to connect with you as an individual. When they use it thoughtfully, everyone wins.

Retail Revolution for Smart Inventory, Customer Insights, and AI Cashiers

Do you ever notice how empty shelves at your favorite store are less common lately? That's not luck, it's AI quietly optimizing what happens behind the scenes. Retailers now use predictive analytics to determine what shoppers will want next week, not just what they wanted last month. Algorithms sift through weather data, social trends, and even local events to help fashion stores avoid running out of popular sizes or colors. If a sudden cold snap is approaching, stores will have winter gear on display before anyone even checks the forecast. These systems don't just react, they anticipate, which changes how retailers handle inventory. Gone are the days of warehouses overflowing with unsold stock or frantic calls to suppliers because the last batch of a trending toy sold out overnight. AI-driven supplier reordering systems take a constant pulse on sales and restock automatically, reducing waste and keeping those "Sorry, we're out!" signs at bay.

Shopping in person or online, you'll see AI making things smoother and, sometimes, almost invisible. Some stores now offer cashierless checkout powered by computer vision. Think of walking into a shop, grabbing your snacks, and simply walking out while your phone buzzes with a receipt. This isn't sci-fi; *Amazon Go* uses a network of cameras and sensors to track what you pick up and charges your account automatically, eliminating the need for waiting in line and avoiding awkward small talk at the register. Online chatbots serve as virtual shopping assistants, guiding you to the correct size or helping you find that elusive deal. They answer questions instantly, help compare products, and can even recommend outfits based on your latest search or purchase. These digital helpers free up human staff for the tasks that still require a person's touch.

Retailers don't just want to sell you something once; they want to stay informed about what you want next. AI delves deeply into customer data to identify patterns that even sharp-eyed managers might miss. By analyzing every click, review, and social media post, businesses can determine which products are poised to trend and which ones are likely to fail. For instance, sentiment analysis tools scan *Twitter* and *Instagram* during a product launch to gauge excitement or spot issues early. Suppose people start raving about a new sneaker style in one city. In that case, stores can ramp up inventory there before demand explodes everywhere else. This isn't just about following the crowd; it's about reading the mood of millions at once, then acting on it before the competition does.

Of course, integrating AI into retail isn't always a straightforward process, especially for companies that have been around since before smartphones were invented. Legacy systems are stubborn. Many supermarket chains have learned to roll out AI in phases rather than implementing it all at once. They might start with automated inventory tracking in a handful of locations before expanding to hundreds more. This way, they catch bugs early without risking major disruptions across their entire operation. Employees often need retraining for new roles, too, maybe learning how to interpret sales dashboards instead of counting

23

boxes or managing customer interactions through tablets rather than registers. Some staff shift from routine checkout work to more engaging problem-solving or customer service roles, which can actually boost morale if handled with respect.

Integrating AI in retail is about blending efficiency with experience. When it works well, shoppers notice fewer headaches, no more sold-out essentials, or endless lines at checkout, and staff are freed up from rote tasks, allowing them actually to help customers in meaningful ways. Retailers who leap thoughtfully end up with smarter shelves, happier shoppers, and teams ready for whatever trend comes next.

Using AI in Finance for Fraud Detection, Robo-Advisors, and Credit Decisions

Picture yourself checking your bank app at lunch, only to see a push alert about a possible fraudulent charge. It could be a weird overseas purchase or a $1 test transaction. Today, those alerts aren't just a result of rigid rules; AI is on the job, monitoring every swipe and transfer in real-time. It constantly learns your usual patterns, comparing them with billions of others. When something unusual occurs, such as a charge from a city you've never visited or a late night ATM withdrawal that's out of character, anomaly detection models flag it instantly. This system doesn't just block random purchases for no reason; instead, it sifts through the noise to identify legitimate transactions. It identifies genuine threats, reducing both losses and those annoying false alarms that leave you stranded at the checkout.

Fraud detection is only one way AI is reshaping finance. The rise of robo-advisors has made wealth management more accessible to a broader range of people than ever. In the past, obtaining investment advice meant scheduling lengthy meetings with a financial planner or attempting to decipher market jargon on your own. Now, platforms like *Betterment* or

Wealthfront ask a few questions about your goals and risk tolerance, then use algorithms to build a portfolio that fits your life, no suit required. These digital advisors rebalance your investments automatically, watch for tax-saving opportunities, and even nudge you to save more without drowning you in charts. For anyone who wants to grow their money but doesn't want to become an expert overnight, this is a game-changer.

AI has also started to rewrite the rules of credit scoring and loan approvals. Traditional credit checks often overlook individuals with thin credit files, those who are young, self-employed, or new to the country. Machine learning models now look far beyond the usual credit report. They analyze rent payments, utility bills, mobile phone usage, and online shopping habits to paint a more comprehensive picture of financial behavior. This means that someone who always pays their phone bill on time but has never taken out a loan can now qualify for credit that would have been previously closed off. Lenders use these models to make faster decisions and offer fairer rates to those who have been overlooked by traditional systems.

Transparency and fairness have become hot topics in this space. When an algorithm decides who gets a loan or flags a transaction as fraud, it's not enough to just say, "The computer says no." Regulators demand explainability, tools that can show why a decision was made in plain English. If your loan application is denied, you should be able to see the key factors, including your income history and spending patterns, rather than just a cryptic rejection letter. Platforms now design explainability dashboards so both customers and regulators can review how models work and catch any unfair bias before it causes harm.

Financial institutions must also walk a tightrope with regulations like the Equal Credit Opportunity Act. Every AI-powered tool needs regular audits to ensure it doesn't discriminate based on race, gender, or other protected traits. Compliance teams double-check that algorithms are trained on data sets reflecting real diversity and that outcomes are consistent across groups. Auditable AI means that every step in the process is logged and reviewable. If something goes wrong, there is a trail showing exactly where and why it happened.

If you've ever worried about a faceless algorithm making decisions about your money, know that consumer protection is front and center in financial AI design. Banks are required to keep you informed, offering clear disclosures and an easy appeal process in case things go wrong. Audits and transparency tools provide peace of mind that these systems aren't just fast, they're fair and accountable. The goal isn't for AI to replace human judgment entirely but to make financial services safer, smarter, and more accessible for everyone trying to build their future.

Education Gets Smarter with AI Grading, Tutoring, and Adaptive Learning

If you've ever waited days for feedback on an essay, you know the frustration of old-school grading. AI is shaking this up, giving teachers a break and students quick, actionable insights. Picture this: you submit an essay online, and within minutes, you get a score plus targeted comments on your argument structure, grammar, and even originality. These systems don't tire halfway through a stack of papers or accidentally skip over a paragraph; they remain consistent. Standardized tests now employ automated essay scoring, where machines read thousands of responses, flag unusual phrasing, and identify patterns that humans might miss. This speeds things up for everyone, allowing teachers to focus on one-on-one help instead of being overwhelmed by paperwork. For students, fast feedback means they can correct mistakes before habits form, rather than discovering the problem weeks later when they barely remember what they wrote.

However, grading isn't the only area where AI steps in. Adaptive learning platforms are revolutionizing how lessons are delivered. Instead of assigning the same worksheet or video to every student, these platforms track student performance in real time and adjust the difficulty or topic on the fly. If someone breezes through algebra basics but stumbles on word problems, the system notices and serves up more practice in that specific area. Math apps for kids now adjust challenge questions based on right or wrong answers, nudging learners past their weak spots while keeping things engaging. Language apps do something similar, switching up exercises if pronunciation needs improvement or introducing more challenging vocabulary as confidence grows. The benefit is personal; students feel seen and challenged at just the right level, not stuck in a one-size-fits-all rut.

AI tutors and virtual teaching assistants are another game changer, especially when teachers' time gets stretched thin. Imagine a chatbot that answers homework questions at midnight or explains a tricky science concept for the third time, without ever losing patience. These digital helpers don't replace teachers but fill in the gaps when extra support is needed. Some universities utilize AI teaching assistants that handle routine student questions about deadlines or course materials, freeing up real instructors to tackle more in-depth discussions. In K-12 settings, chatbots help keep kids on track with reminders and encouragement. At the same time, parents gain peace of mind knowing help is available outside of school hours.

Accessibility is where AI shines brightest. Students who once struggled to keep up now have tools that meet them where they are, allowing them to succeed. Speech-to-text software helps learners with dyslexia write essays by allowing them to speak instead of typing. Real-time translation engines enable kids from different language backgrounds to participate fully in class discussions, breaking down barriers that previously felt insurmountable. AI-driven reading aids adjust the font size and contrast or even read text aloud, making materials usable for those with visual impairments or learning difficulties. For students who face anxiety about speaking up in class, chatbots provide a low-pressure way to practice language skills before stepping into the spotlight.

It's not just about fancy gadgets or showcasing what's possible; it's about genuine inclusion and allowing every learner to succeed. AI can detect subtle learning patterns that traditional methods miss, such as a student who consistently aced multiple-choice questions but freezes during short-answer sections, or someone who absorbs spoken instructions but tunes out during lectures. By tracking granular data across assignments and interactions, AI identifies trends and flags students who might be quietly falling behind, allowing educators to intervene early rather than after the final grades are posted.

Education powered by AI doesn't erase the human touch; it amplifies it, allowing teachers to focus on creativity, mentorship, and building genuine relationships while machines handle repetitive tasks. The result? Classrooms that adapt on the fly support every learner's needs freeing up educators to do what they do best: inspire curiosity and growth.

AI in Creative Industries from Generative Art to Automated Copywriting

Creativity is undergoing a radical transformation, and it's happening right before our eyes. Whether you're a graphic designer, a copywriter, or just someone who likes to experiment with visuals for your side hustle, AI has kicked open the doors to new ways of making and sharing ideas. Artists and designers are using tools like *DALL-E* and *Midjourney* to whip up jaw-dropping images, sometimes from nothing more than a quirky prompt typed into a box. You could ask for "a sunset over a city in the style of Van Gogh," and seconds later, you're staring at a unique piece of art that never existed before. These platforms don't just spit out clones; they remix, mash-up, and surprise, pushing creative boundaries in ways that would take humans hours or even days.

Writers aren't left behind, either. Blog posts, ad copy, and newsletters, AI helps get them started or polishes drafts in record time. *ChatGPT* has become the go-to for busy marketers and content creators who need a nudge past writer's block or want to brainstorm headlines that catch attention. You can feed it your rough ideas or even just a few keywords, and it returns suggestions that save precious hours. Instead of staring at a blank page, you're suddenly editing and refining, allowing you to focus on what matters: your message and your unique style.

But it's not just about solo artists or lone writers. The real magic happens when humans work in tandem with machines. Filmmakers are already utilizing AI-driven editing tools to suggest alternative video cuts for *YouTube* or *TikTok*, sometimes identifying patterns or beats that even seasoned professionals might overlook. A creator might upload raw footage, and the AI analyzes pacing, visual interest, and even trending styles to recommend fast edits or transitions. This frees up creative energy for storytelling while reducing repetitive, mundane tasks. Musicians now use AI to generate backing tracks or harmonies, enabling them to explore fresh ideas without hiring a full band. The tech doesn't steal the spotlight; it acts as a superpowered assistant, amplifying what people do best.

Of course, with all this power comes a complex web of questions about ownership and credit. If an AI makes music that gets millions of streams, who owns the copyright, the person who wrote the prompt, or the company that trained the AI? Legal disputes have already made headlines, particularly in the realm of music and visual art created by algorithms trained on existing works without explicit permission. Some organizations now establish guidelines for transparency, requiring creators to disclose when AI has played a significant role in a project. This helps audiences understand the distinction between what is human-made and what is machine-assisted, an important distinction, as the lines between inspiration and automation become increasingly blurred.

For small business owners, freelancers, and influencers, AI has lowered the barriers that previously kept creative projects out of reach. Imagine needing a logo but not having the budget for a designer; AI logo generators can create dozens of options in minutes based on your input. Social media creators use video editing apps powered by machine learning to tweak color grades, cut footage to music, or add subtitles, all with just a few taps on their phones. What used to require expensive software and years of training now feels accessible to anyone with an internet connection and a sense of curiosity.

The democratization of creativity means you don't need fancy credentials to make something new and share it with the world. People with no design background can launch compelling brands. At the same time, those who used to struggle with words find their voice through AI-assisted copywriting. This doesn't erase the need for taste, vision, or editing skills; it makes those qualities more critical. The secret is knowing how to guide these tools so they reflect your personality and goals rather than just churning out generic work.

As we wrap up this chapter, one thing stands out: AI isn't here to replace creativity; it's here to expand it. Whether you want to make art, write stories, build a brand, or solve everyday problems with flair, the tools are more accessible than ever. In the next chapter, we'll shift gears from creative work to practical AI projects you can try right away, no prior experience required.

Chapter Three

Getting Hands-On Immediate, Practical AI Wins for Your Workflow

A Beginner's Evaluation Checklist for How to Pick Your First AI Tool

Feeling lost in the sea of so-called "AI-powered" tools? You're not alone. It can be as overwhelming as walking into a giant electronics store where every device promises to change your life, but few deliver on their promises. The key is to find an AI tool that meets your needs and doesn't become another unused app. Here's a straightforward checklist, no tech expertise needed, to help you confidently choose your first AI tool and avoid the hype. Understanding these tools will empower you and give you a sense of control over your workflow.

First, clarify your goal for the tool, even if it's just to "save me time on boring stuff." That's a perfect starting point. Next, check how easy it is to use. The best entry-level AI tools are usually labeled "no-code" or "low-code." No-code tools require no programming, just basic clicking, dragging, or filling in forms; low-code tools might ask you to set a few rules, but nothing advanced. If the interface is packed with code, look elsewhere. For most beginners, no-code tools are an ideal choice.

Another key factor to consider is integration. Consider your current workflow: do you primarily use *Google Workspace*, *Slack*, *Trello*, or a combination of these? The best AI tools are those that seamlessly integrate with the apps you're already using. Look for direct connections; for instance, some content generators work within your document editor or social media dashboard, while meeting assistants can sync with your calendar and video calls. 'API integration' is a good sign, indicating that the tool can easily connect with many other apps. If you're a mobile user, ensure there's a reliable mobile app or a responsive website for on-the-go access.

Focus on features that matter:

- **Privacy:** If the tool accesses your email, files, or contacts, verify that it has a clear and easily accessible privacy policy. It should clearly outline what data is stored, who has access to it, and how it's safeguarded. Avoid tools with vague or hard-to-find privacy info.

- **Customer support:** You'll want quick help if you get confused or encounter an issue. Look for live chat, forums, or, at the very least, a responsive support email.

Trial periods and freemium models are great for getting hands-on experience without risk. A free trial allows you to test a product before making a purchase; a freemium tool offers core features at no cost while charging for additional options. Use these trials to determine how well a tool fits into your daily routine and whether it saves you time and effort.

Watch for deal-breakers. Be wary of tools making wild promises ("100% automated creativity," "one-click revolution for your business," etc.). Real AI helps you work smarter, but it's not magic. Avoid tools whose decision-making process is a black box; if you can't tweak settings or understand why the AI makes its decisions, move on. Practical tools enable you to adjust inputs and settings to achieve outputs that align with your intentions. Remember, over-automation can be a pitfall, so stay cautious and mindful.

Ethics and transparency count. Steer clear of tools that are vague about data use or lack transparency in their operations. Trustworthy AI companies offer plain-language transparency reports and explain their training methods. They should give you options to opt out of data collection.

To make comparison easy, use this worksheet:

Your AI Tool Evaluation Worksheet

1. Name of Tool:
2. What problem does this solve for me?
3. No-code/Low-code? (circle one):
4. Integrates with (list apps):
5. Is the Privacy policy easy to find and read? (Y/N):
6. Customer support available? (Live chat / Email / Forum):
7. Free trial/freemium? How long?:
8. Any marketing promises that feel too good to be true?:
9. Can I understand or adjust how the AI works? (Y/N):

10. Would I trust this tool with sensitive info? Why/why not?:

Sample Use: Social Media Content Generator

- *Jasper*: No-code, integrates with *Google Docs* and *Canva*, clear privacy policy, live chat, 7-day trial, promises "brand voice consistency," allows manual edits after generation.

Sample Use: Meeting Assistant

- Example Tool: *Low-code*, syncs with *Outlook* and *Zoom*, includes privacy information in setup, offers email support, and is free for 10 meetings per month. It claims to provide "instant insights," allowing users to review and edit summaries.

Fill out the worksheet as you try new tools and compare a few before making a choice. Don't rush; finding the right AI should make life easier, not more complicated.

Red flags:

- If the privacy policy is unclear or complicated to find, skip it.

- Avoid any tool labeled "AI-powered" that's just basic automation. True AI gets better with use.

- If you can't provide feedback or get help quickly, look elsewhere.

Choosing your first AI tool can be a stress-free and straightforward process. With this checklist and a bit of curiosity, you'll find something that fits your workflow and makes repetitive tasks less tedious. The relief and hope that come with finding the right AI tool are just around the corner.

Streamlining Meetings with AI Note-Takers and Summarizers

If you're tired of blank documents or hunting for follow-ups after meetings, you're not alone; meetings often become chaotic without clear notes or tracked action items. AI note-takers and summarizers are transforming this by providing organization and clarity. Tools like *Otter.ai* transcribe conversations in real-time, capturing speech, tone, and context and generating summaries so you don't miss important points even if you momentarily tune out. *Fireflies.ai* takes it a step further by tagging action items, decisions, and questions, making post-meeting follow-ups straightforward and efficient. *Zoom's* built-in AI summaries auto-generate recaps post-call, highlighting meaningful discussions for everyone.

These tools are designed for easy setup and use. Most connect directly to your existing calendar and meeting platforms, such as *Google Calendar*, *Outlook*, *Zoom*, or *Teams*, so there's minimal friction. After signing up, you link your calendar and choose permissions for note access and recording storage (cloud or local). Many let you set rules, such as only recording specific meetings or attendees. For new users, try a practice session to become familiar with the interface and privacy controls.

Teams quickly notice the difference. Sales groups, for instance, used to scramble for notes and often forgot details or missed follow-ups. With AI note-takers, everyone receives clear summaries and action items right in their inboxes. This speeds up responses, reduces the need for reviewing recordings, and ensures that nothing is missed. Remote teams especially benefit; searchable transcripts and tagged action items help everyone stay informed, even if someone misses a meeting.

Respect privacy and meeting etiquette with AI. Always inform attendees if you're using a note-taking bot ("We're using Otter.ai today to capture our discussion"), which builds trust. For sensitive topics, many tools allow users to pause recordings or restrict access to transcripts. Be aware of your company's policies regarding transcript storage and access. Transparency and honor for opt-outs minimize the risk of confusion or legal issues.

AI note-takers save time, improve accountability, and reduce meeting stress. With better records of what's discussed, clear ownership of follow-ups, and automated reminders, meetings become more productive, with less confusion and wasted time.

AI-Powered Email Triage and Smart Scheduling for Busy Pros

Waking up to an overflowing inbox packed with urgent, random, and never-ending messages can burn hours from your day. AI-powered email prioritization changes this by filtering your inbox so you only see what matters. With *Gmail's "Priority Inbox,"* AI analyzes your behavior, including which emails you open, ignore, reply to, or delete and how quickly you do so. It considers the sender, your response times, key times of day, and keyword patterns. The result? Important emails surface at the top, while less relevant ones quietly drift out of sight. The system's learning capabilities refine over time, saving you effort and freeing up your focus for meaningful work.

For those needing more speed and precision, tools like *Superhuman's AI triage* offer powerful automation. Messages are sorted into categories like urgent, follow-up, newsletters, or "later." Critical threads are flagged, and conversation summaries spare you from clicking through endless replies. Your inbox appears as a clear, organized dashboard, panic replaced by clarity. Bite-sized previews enable you to determine what requires your attention quickly, and the tool sends reminders if you miss key deadlines or replies, serving as a digital safeguard.

Finding mutually convenient meeting times used to mean endless email chains ("Does Tuesday at 3 work?" "How about Wednesday?"). AI scheduling tools, such as *Calendly*, eliminate that chaos. By connecting calendars, they suggest open slots for all parties. You send a link, others select what fits, no more double-bookings or time zone slip-ups. For more complex scheduling, tools like *x.ai* act as a virtual assistant that coordinates on your behalf negotiate across different calendars, and inserts buffer breaks to protect against back-to-back meeting fatigue.

Setup is easy. Connect your email and calendar accounts; most tools support *Outlook* and *Google Calendar*. Grant permission to access events (and sometimes emails) with customizable privacy controls. Adjust your alert preferences to receive reminders before meetings or a daily summary of your priorities. Good tools let you pick whether you're notified of every change or just given an end-of-day digest.

You retain complete control with these systems; AI is intelligent but not perfect. Occasionally, it will mark a newsletter as high-priority or miss a key message buried in a thread. Please take a minute each week to adjust the rules: flag important senders or subjects, teach the system to treat particular items (such as newsletter blasts or reports) as low-priority, and continually improve its accuracy.

Automation works best with clear boundaries. Block out lunch or personal time so bots never book over them, and set up focus time so colleagues know not to disturb you. If you chronically accept meetings back-to-back, program the tool to add buffer slots for sanity and short breaks.

If an AI schedules something at an inconvenient time, during a personal commitment, for example, you can always override it. The bot works for you, and manual selection is always available as a backup.

Double bookings once plagued manual scheduling, especially with separate personal and work calendars. Now, AI assistants scan all your authorized calendars to prevent conflicts, and "do not disturb" hours or recurring out-of-office blocks automatically sync across all devices.

A helpful tip:

Check the settings or audit panel occasionally to review how emails and contacts are being categorized and retrain or adjust preferences as needed. Most platforms allow reclassification without losing data.

AI-powered tools strike a balance between structure and flexibility, streamlining communications and scheduling based on your habits while never restricting you. As these systems learn from your tweaks, their ability to distinguish urgently actionable items from low-priority noise constantly improves, leaving you in control amid the digital swirl.

If you're new, start simple: activate *Gmail's Priority Inbox* or try a free scheduling tool. Connect main accounts, test out its sorting with non-sensitive emails, and experiment with scheduling casual appointments. Fine-tune rules for key contacts, such as your boss, partner, or major clients, and you'll notice an improvement in your efficiency.

The benefits add up: less noise, quicker decisions, fewer missed messages or meetings, and more time for work or a quiet cup of coffee before the next onslaught.

Setup and Pitfalls of Using Chatbots for Customer Service

Choosing your first chatbot platform is like hiring a tireless team member who's always available. With numerous options available, beginners require something practical, flexible, and not overly complex. *Intercom* is a strong contender for website chat, requiring just a quick code snippet to get started. *Drift* is also popular, particularly for qualifying leads by asking targeted questions and directing visitors to the right resource or person. Both platforms offer user-friendly drag-and-drop builders, so you don't need programming skills. Look for tools that provide analytics dashboards, thorough documentation, and a sandbox mode for testing your bot before going live.

Once the bot is set up, customization is key. Out-of-the-box bots are generic and often ignored by visitors. To maximize the effectiveness of your bot, tailor its scripts to align with your company's tone and address the most frequently asked questions by your team. Start with the five questions you receive often, such as shipping information, returns, troubleshooting, or appointment bookings. Craft concise, conversational responses and let your bot handle these automatically. Most platforms support uploading FAQs, and after launch, reviewing bot logs exposes commonly asked questions, allowing you to refine its content continually. Honest feedback, such as users asking about features you haven't covered or finding confusing responses, is invaluable for training your bot to become more helpful.

AI learning features amplify chatbot performance. Tools like *Drift* and *Intercom* can analyze chat data to improve future responses. If users frequently rephrase questions or leave after receiving specific answers, that signals an area that needs refinement. Some platforms even flag confusing conversations and suggest better phrasing. The more you review and revise the chatbot based on actual usage, the more effective it becomes at handling routine requests, allowing your staff to focus on complex, high-value issues.

Measuring your chatbot's impact isn't just about hoping for fewer angry emails; it's about tracking specific data. Launching a bot should drastically reduce your first-response time, making customers feel recognized and assisted instantly. Review the ratio of queries resolved by the bot (self-service rate) versus those sent to humans (escalation rate). A high escalation rate isn't always alarming; it means the bot recognizes its limits and doesn't frustrate users by getting stuck in endless loops. Customer satisfaction surveys after bot chats ("Was this helpful?") provide critical feedback, and some tools can even detect frustration signals, such as repeated agent requests or negative sentiments.

It's tempting to automate everything, but that can frustrate customers with robotic, unhelpful answers. Avoid over-automation by including easy ways for users to request a human after a couple of failed attempts. Don't neglect accessibility; ensure your chatbot works with screen readers, supports keyboard navigation, and avoids flashy graphics or tiny fonts. Regularly test accessibility using simple tools or customer feedback to ensure optimal usability.

Another frequent pitfall is neglecting to update content. Products, policies, and promotions change frequently, so keep your chatbot's information up to date with monthly content reviews. Many platforms allow you to add expiration tags or review reminders for responses, so nothing becomes outdated.

Treating your chatbot as a "set-it-and-forget-it" tool is a risky approach. Customer expectations evolve as they interact with more intelligent bots elsewhere, so regular updates based on actual conversations are essential. Ask your support team to flag unclear exchanges or new questions so the bot's knowledge base stays fresh and valuable.

Finally, evaluate success by more than just ticket volume. Examine whether customer satisfaction improves, repeat contacts decrease, and agents have more time to address nuanced issues rather than repetitive ones, such as checking orders. When your chatbot frees up staff for more complex tasks and reduces customer wait times, it delivers real value.

In summary, service chatbots should deliver quick answers to users, reduce repetitive tasks for staff, and provide valuable insights for business leaders. The most effective bots are continually refined, have clear escalation paths, and adhere to accessibility from the start. Treat your chatbot as a vital, evolving part of the customer experience, not just another website widget, and you'll see ongoing benefits for both your business and your team.

Automating Repetitive Workflows with No-Code AI Tools

You're overwhelmed with repetitive tasks, such as shuffling files between folders, manually copying information from one app to another, or tracking down receipts for invoicing. Now, imagine handing these over to a digital assistant who never gets tired or distracted. That's what "no-code" AI brings to your table. In plain terms, no code means you don't have to write a single line of programming to set up intelligent automation. Everything is visual: drag and drop, click, and customize. It's like building with digital *Lego* blocks instead of piecing together complicated blueprints. This shift is massive for anyone who isn't a developer, as it puts powerful automation within reach for busy individuals who want results quickly.

Zapier is a standout here, acting as the ultimate connector for your favorite apps. Say you get invoices as email attachments; with *Zapier*, you can create an automation that detects each incoming invoice, saves the attachment to *Google Drive*, and then logs the transaction in a spreadsheet. No more manual downloads or lost files. Scheduling social posts is another great fit; link a shared *Google Sheet* with planned content to your social media accounts, and *Zapier* will post updates automatically based on your schedule. *Microsoft Power Automate* works similarly but is better suited for business processes within organizations. For example, it can take survey results from *Microsoft Forms* and instantly update a *SharePoint* database or trigger alerts in *Teams* when a new entry is received. The beauty is that these tools work quietly in the background; once set up, they just run.

Setting up these automations is straightforward. Start by selecting a trigger, the event that triggers everything. Maybe it's receiving an email with "Invoice" in the subject line or updating a row in your spreadsheet. Next, define the actions: save a file, send a notification, create a calendar event, or update a Customer Resource Management (CRM) record. Each step is clearly explained in plain language, and a preview button is usually available to test before going live. In under an hour, even if you've never automated anything in your life, you can have workflows humming along.

Different professions benefit from tailored templates that save even more time. HR teams can automate resume screening: every new application received via email gets logged in a tracker, sorted by keyword matches, and flagged for review, all without touching your inbox. Marketing teams often set up lead capture flows. When someone fills out a form on your website, their info is instantly added to your CRM and triggers a personalized follow-up email. Operations managers can utilize automation to monitor inventory. Suppose a supplier sends an email indicating low stock on a critical item. In that case, Power Automate can create an alert in Slack or Teams, allowing your team to react before a shortage occurs.

Scaling up these automations is easy with the right approach. As you add more workflows, monitoring becomes essential. Most no-code AI platforms feature dashboards that display recent activity and any errors that occur. Set up notifications for failed automation so you know right away if something breaks, such as a missed invoice upload or a skipped message, so that these issues don't go unnoticed. Reviewing logs every week highlights recurring problems (such as authentication errors or missing data). It helps you fine-tune triggers or steps for improved reliability.

Troubleshooting doesn't have to be stressful. Most issues are simple: you updated your password. You forgot to refresh the connection, or the format of incoming data changed slightly. Step through the automation's logs to see exactly where it went off track. Suppose problems persist or you encounter more complex needs, such as integrating custom APIs or handling thousands of records simultaneously. In that case, it may be time to seek IT assistance or explore premium support options from the platform.

No-code AI isn't just about saving time; it's about freeing up energy for more innovative thinking and creative work. You can skip tedious steps and focus on what moves projects forward. It's empowering and surprisingly fun to see your digital workflow clicking along, with notifications pinging at just the right moment or files landing perfectly where they should.

Workflow Brainstorm: Quick Wins for Your Role

- **HR:** Automatically log new candidate emails to Trello or Asana with attachments.

- **Marketing:** Sync new leads from Facebook ads into Mailchimp for instant outreach.

- **Sales:** Alert the team in Slack when contracts are signed via DocuSign.

- **Finance:** Generate weekly expense reports from Google Sheets updates.

- **Operations:** Flag supplier delays from emails and updates tracking spreadsheets instantly.

Automations like these don't just save minutes; they add up to real hours every week, which you can spend on strategy, growth, or even just taking a proper lunch break.

As we wrap up this chapter, keep in mind that no-code AI turns everyday routines into efficient systems that work for you, not against you. With each workflow you automate, you chip away at busy work and create space for more in-depth work and better results. Up next, we'll look at how AI can fuel growth in your career and help you learn new skills, so the benefits go way beyond streamlining tasks.

Chapter Four

Upskill Paths and Choosing-Your-Own-Adventure Learning in AI for Career Growth

Mapping Your AI Learning Journey with Zero Prerequisites and Real Outcomes

Perhaps you've seen colleagues advance with AI or noticed job postings that are loaded with "artificial intelligence" and "automation." It's common to feel hesitant, especially with new jargon everywhere and no clear entry point. But you don't need a computer science background or math skills to find real results and shape your AI journey. The key is a step-by-step plan tailored to your current situation and your desired outcome, with flexibility built in along the way, empowering you to take control of your learning journey.

Let's break down potential paths based on your starting point and desired outcome. If you're not technical, if *Excel* feels intuitive but *Python* is a mystery, your journey is "From AI-curious to AI-confident." Begin by learning what AI can do in your field: read a couple of easy explainers, then experiment with no-code tools through guided demos. Many platforms offer beginner-friendly projects, like automating a basic task or reviewing feedback data. Dedicate just ten minutes daily to micro-challenges, such as using an AI email assistant or setting up a simple chatbot at work. Within a week or two, the language and concepts, such as "model," "workflow," and "automation," will begin to feel familiar. The more hands-on you get, the faster your confidence will grow.

If you're already skilled with spreadsheets or basic databases, your route is "From *Excel* power user to AI workflow builder." Go beyond formulas and pivot tables to tools that automate tasks or reveal patterns in larger datasets. For instance, automate a weekly report or connect two apps using an AI-powered integration to build digital shortcuts that reclaim hours each month. Learn best by doing: use interactive tutorials to develop and tweak workflows. Set small, clear goals, like "automate my team's data collection" or "summarize customer reviews automatically," and acknowledge each success.

Regardless of your background, setting clear and achievable goals is pivotal. Avoid vague objectives, such as 'learn AI.' Instead, be specific: 'Automate meeting scheduling by next month' or 'Explain machine learning to my team without stumbling.' Break these big goals into smaller, manageable micro-goals. For instance, if you aim to curate content with AI, test out three tools and find the one you like best. If your goal is to communicate effectively with AI at work, identify three frequently asked questions and practice explaining them clearly. Remember to celebrate each milestone, no matter how small.

When embarking on your AI learning journey, it's crucial to choose practical, accessible starting points. Free or low-cost options that fit into your busy life and don't require marathon study sessions are ideal. Podcasts offer bite-sized learning during commutes (see *Best AI newsletters and podcasts*, 2025), while short video series on *YouTube* or *LinkedIn* Learning help you see concepts in action. Even ten-minute daily tasks, such as teaching AI to filter spam or automating document sorting, can build momentum. Remember, the focus should be on progress, not perfection.

Expect setbacks. Everyone encounters them, whether it's feeling not technical enough, losing focus, or getting anxious about complex math. Most real-world AI tools hide the technical details behind user-friendly interfaces. If you worry about messing up or not understanding, remember that even experts make mistakes when learning. It's perfectly fine to Google basic questions or ask "simple" questions in online groups. You might help others by doing so.

Roadblock Busters and Motivation Boosters

Think about the last time you learned something new, such as a guitar, a recipe, or public speaking. What helped you stick with it? Jot down three things that supported your progress (like joining a community, setting small goals, or rewarding yourself). Apply those lessons to your AI journey: join a supportive *Slack* group (see *Top 13 data science and machine learning Slack communities*, 2024), where you'll find reassurance and advice, set a weekly challenge (such as automating a single task), and recognize all achievements, big or small.

Here's a story for perspective: Amy used to think she'd never get AI after struggling with high school math. She began with ten-minute tutorials on automating her inbox and schedule. Within a month, she built a no-code workflow to sort customer requests. Her confidence soared, leading her to larger projects and, eventually, a digital transformation leadership role. Her tip: Celebrate each small win, and don't stress about what you haven't learned yet. Remember, every step forward is a step towards your AI proficiency.

Tracking your progress matters. Keep a simple checklist: Did you tackle today's micro-challenge? Have you tried explaining an AI concept to someone else? Each check-in builds momentum. Set reminders to review your progress every few weeks. You'll likely be impressed by how far you've come.

Regardless of your tech background, there's an AI path for you. Whether you start with a phone app or launch a major automation at work, take the first step and let curiosity guide you.

AI Upskilling for Marketers, Managers, and Consultants

AI skills now vary widely by profession. For marketers, key strengths include prompt engineering and campaign optimization. Crafting effective prompts enables you to generate copy, headlines, and social posts tailored to your audience in minutes, allowing for rapid iteration and A/B testing. Tools like *HubSpot's* AI features enable advanced segmentation for smarter retargeting and audience discovery; tasks that once required hours of guesswork are now achieved with a few clicks and are immediately evident in performance metrics.

Managers benefit by automating reporting and utilizing AI-powered analytics. Familiarity with *Tableau* or *Power BI* provides instant access to dynamic dashboards built from live data, enabling early detection of trends and automated forecasting, with no deep technical skills required. Recurring reports become automated, freeing up time and providing on-demand insights for meetings and planning.

For consultants, value comes from assessing clients' AI readiness and comparing vendors to determine the best fit. Clients seek credible advice on which AI tools deliver a return on investment (ROI). Your expertise lies in placing AI into workflows, vetting vendor claims, and translating technical features into business value. Utilize frameworks for solution evaluation, basic comparison sheets, and compliance checklists to advise clients with confidence, eliminating the need for coding.

Hands-on experience matters for all. Marketers can use Jasper for generating creative content and HubSpot for analytics. Try building, testing, and refining email sequences for real customer segments to showcase their impact. Managers can automate recurring dashboards using *Tableau* or *Power BI* to streamline weekly or onboarding reports. Consultants should utilize vendor analysis templates and workflow automation maps to identify opportunities for efficiency, accompanied by tangible ROI projections.

Projects that demonstrate real impact stand out. Marketers should document AI-driven campaigns that improve open or conversion rates, using before-and-after snapshots as proof. Managers can showcase dashboards that decrease reporting times or reveal cost-saving insights. Consultants should highlight client cases where their recommendations led to better outcomes or efficiency.

Quantifying results boost career prospects. Instead of just listing skills, illustrate impact: "AI-powered targeting increased CTR from 2% to 5%," or "automated dashboards saved four hours a week." Pair metrics with brief stories, such as "After automating onboarding, client referrals rose by 18%," and tie these outcomes to personal career advancements.

Make your AI projects visible. Post them on LinkedIn or in relevant online groups. Join *Slack* communities to share work, get feedback, and find collaborators. The more visible your skills, the more opportunities come your way.

Stories from Career Changers for Pivoting to an AI-Driven Role

Transitioning into AI-powered roles isn't just for techies or coders. Professionals from many backgrounds are making the switch and discovering that their previous skills are highly valuable. For instance, Priya, who taught high school science for a decade, grew tired of grading but wanted to continue making an impact. Instead of moving to another classroom, she explored edtech through webinars and online forums, experimenting with AI education tools. She eventually became an AI product evangelist at an edtech startup, using her classroom insights to provide valuable feedback to designers and engineers. Her communication and empathy skills enabled her to connect with both customers and tech teams, acting as a vital bridge.

Similarly, Greg, a longtime retail manager, dealt daily with inventory stress and late shipments. When his company adopted AI-driven inventory management, Greg co-led the rollout, learning how to analyze AI dashboards despite his background in people-focused roles. Witnessing the efficiency AI brought, he networked with other managers and joined workshops. Within a year, he transitioned into supply chain analytics, where his on-the-ground retail experience provided him with perspectives that pure data specialists often overlooked.

What enabled these shifts? Not advanced coding, but transferable skills: problem-solving, adaptability, clear communication, critical thinking, and project management. Priya's empathy, sharpened by her teaching experience, enhanced her ability to understand user needs. At the same time, Greg's unflappability enabled him to handle the pressure of implementing new systems. Both relied on their organizational habits, including running meetings, managing tasks, building consensus, and strategic questioning. In many AI-adjacent jobs, these "soft" skills are just as crucial as technical know-how, especially when bridging the gap between users and tech teams.

Career pivots often begin with a personal assessment. Automation and digital transformation pressure mid-career professionals to rethink their paths. Priya noticed AI tools spreading in schools and realized her job would eventually evolve. Acting early, she identified what she loved in education as she networked with those already in edtech and sought out relevant projects. Greg's path began with curiosity and evolved into an opportunity as he experienced firsthand the benefits of AI systems.

If you're considering a move, map out your motivations and readiness before making a leap. Ask yourself:

- Is your industry undergoing tech-driven changes?

- Do you enjoy learning new systems or tools?

- Can you work with technical teams and explain complex ideas in a simple way?

- Are you willing to spend some evenings or weekends learning?

- Does your network include anyone in AI-adjacent roles?

If you principally answer "yes," it could be time to explore further.

Practical steps help when you're ready to start. Schedule at least three informational interviews with professionals in interesting roles. Don't focus solely on job titles; ask about their daily tasks, the challenges they faced after transitioning to AI-driven work, and the essential non-technical skills they developed. Sample questions include:

- "What surprised you most about your transition?"

- "What do you wish you'd learned sooner?"

- "How does your old role help now?"

- "What do hiring managers value besides technical skills?"

- "Which communities or courses do you recommend?"

- "What's a common mistake newcomers make?"

Build a six-month transition plan:

- **Month one**: research target industries and roles, and assess your skills and gaps.

- **Month two**: join two AI-focused Slack or Discord communities.

- **Month three**: shadow someone in your desired role or volunteer for an AI-related internal project.

- **Month four**: complete a relevant online course, document your learning, and share small wins on LinkedIn or in your community.

- **Month five**: conduct at least three informational interviews and update your resume with AI-related projects (even if internal or side projects).

- **Month six**: apply for entry-level or adjacent roles, leveraging both your expertise and new skills.

Common mistakes include undervaluing your current skills, thinking you need to become a full developer immediately, or searching only for generic "AI jobs" instead of roles that combine your background with technology, such as AI operations, implementation lead, or customer success manager for innovative products. Don't neglect networking;

conversations can reveal opportunities that are often hidden from job boards.

You don't need all the answers at the start. Testing small projects, building cross-industry relationships, and connecting with mentors will help you navigate your unique path, one that fits both your strengths and the future of work.

Showcasing Real Projects for Building Your AI Portfolio

If you want to stand out in today's job market, you need proof that you can use AI for real, hands-on work that goes beyond buzzwords on a resume. Experience sticks more than theory, and nothing says "I can do this" like a project you built yourself. It doesn't matter if you're aiming for your first AI-adjacent role or want to be the go-to person at work when someone asks, "Who knows this stuff?" Employers, clients, and even colleagues want to see practical outcomes. That's where a proper AI portfolio comes in.

Start by picking projects that feel real and relatable. Perhaps you've noticed that your favorite local café always has a line at the register and wonder if a chatbot could help with online orders or address frequently asked questions. Building a simple chatbot for them, even as a volunteer, demonstrates that you can identify a business problem and solve it with technology. Or perhaps you're always the one who gets stuck sending out weekly sales or inventory reports. Automate that task using a tool like Zapier, Google Sheets, or Power BI, then document how much time you saved and how it made your team's life easier. These stories pack a punch because they demonstrate your impact in a way that anyone can understand.

The way you present these projects matters just as much as the projects themselves. Don't just list tools used or technical steps. Frame your work around the problem you tackled and the difference it made. For example, say you built a chatbot: explain what the business struggled with before (maybe slow response times or missed messages), what the AI solution changed (customers got instant answers 24/7), and, most importantly, back it up with numbers. Did the café get more online orders? Did customers leave better reviews? If you've automated a report, share the before-and-after statistics: maybe it used to take two hours every Friday, but now it runs in five minutes. Revenue gained, time reclaimed, errors reduced—these concrete wins bring your portfolio to life.

A great project story follows a simple pattern:

Problem → Solution → Outcome

Start with why the project mattered. Move on to what you built, describe your process, but keep it jargon-free so anyone can follow along. Wrap up with the results, especially anything that can be measured. If you faced setbacks (and everyone does), include what went wrong and how you adjusted. That honesty shows resilience and a fundamental understanding of how things work in practice.

When it comes to showcasing your projects online, clarity and structure always prevail over fancy design. A basic personal website, or even a well-organized *LinkedIn* section, does the trick. Organize your content so that readers can see your skills at a glance. Use an outline like this:

- **About:** Who you are, what drives your interest in AI, what problems you love solving.

- **Projects:** A list of your best work, not every experiment or half-finished idea, but two or three solid examples that fit jobs or clients you want next.

- **Outcomes:** For each project, include real numbers and feedback wherever possible.

- **Reflections:** What did you learn? How would you improve next time?

- A few clear screenshots or videos go a long way; seeing your chatbot live on a business's website or an email showing an automated report in action is powerful.

One of the biggest mistakes I see is "portfolio inflation." That's when people add every tiny project they ever touched, even if they're all the same or don't fit their target role. Quality beats quantity. A portfolio with three strong, diverse projects will always impress more than ten nearly identical chatbots or endless lists of basic automation. When reviewing your work, ask yourself: "If I were hiring for this job, would this project catch my eye?" If it doesn't align with where you want to go, either update it or leave it out.

Regularly update your portfolio as your skills grow. Technology changes rapidly; last year's trendy tool may look dated by next month. Whenever you complete a new course, test out a fresh tool at work, or run a pilot project for a friend's business, add it to your site. Set a reminder every few months to review your projects and make sure they still reflect your best self.

If something didn't go as planned, such as a chatbot that flopped or an automation that broke after launch, don't hide it. Instead, write up what happened and what you learned. The chatbot couldn't answer specific customer questions and required better training data. Your automation saved time, but it also missed crucial exceptions that nobody expected. Employers appreciate individuals who take responsibility for their mistakes and demonstrate growth.

Is Your Portfolio Relevant?

- Does each project clearly state the business problem solved?

- Are the outcomes measurable (time saved, money earned, error rate dropped)?

- Is your work presented in plain language, with visuals where possible?

- Have you included recent tools or methods?

- Did you explain lessons learned, even from failure?

- Would these projects interest someone hiring for the roles you're looking for?

Keep these questions handy when reviewing or adding to your portfolio. Being honest about your skills and focusing on what's relevant will help your work stand out for all the right reasons.

Certifications, Courses, and Communities That Deliver

Finding a certification that means something on your resume is like searching for that perfect pair of jeans. There are a million out there, but only a handful fit just right. When it comes to AI, the *Google AI Professional Certificate* stands out for both newcomers and individuals seeking hands-on skills. This one is recognized by employers and covers practical basics, walking you through projects you can actually use at work. If you want broad, big-picture understanding without getting buried in code, Coursera's "AI for Everyone" nails it. Business leaders, non-techies, and even creative professionals recommend this course for its clear breakdowns and real-life examples. Both certificates offer flexible pacing, allowing you to fit lessons into lunch breaks or complete modules over weekends. For those looking to future-proof their careers with formal credentials, these two consistently come up in recruiter conversations and job listings.

Digging deeper into online courses reveals that several platforms raise the bar. Udacity's nanodegree programs have a reputation for being rigorous, up-to-date, and loaded with practical content. The projects aren't just theoretical; they focus on skills that employers want right now, such as building AI models or deploying chatbots. Instructors have real-world industry experience, so you're learning from people who have shipped products, not just researchers or academics. Peer reviews add another layer of value, allowing you to see how others approach similar challenges and providing you with honest feedback on your progress. MIT Sloan's AI management program is a distinct offering designed for leaders and decision-makers who must steer teams and strategy in an increasingly AI-driven world. The focus is on case studies, frameworks for evaluating return on investment (ROI), and how to make ethical decisions regarding automation and data use. There's a mix of live sessions and self-paced modules, plus access to a network of ambitious professionals. Suppose you're eyeing senior roles or plan to influence tech adoption within your organization. In that case, this course provides you with the language and credibility that stand out.

The right community can also be a game-changer. You don't have to go it alone or get lost in endless Reddit threads when you hit a wall. The *Women in AI* global network is a supportive and inclusive space filled with resources for learning, mentoring, and career progression. Whether you're looking for your first AI job or want to connect with other founders, this group brings together a diverse membership from around the world. *Data Science Society Slack* is another favorite, a buzzing spot for swapping code snippets, asking questions, sharing resources, or even hunting for job leads. You'll find channels dedicated to everything from deep learning to data storytelling. There's always someone online willing to review your project or help troubleshoot. Local meetups and *LinkedIn* groups add an offline dimension; nothing beats meeting others at coffee shops or panel events, where you can build relationships and learn what's working in your city or industry.

Staying current isn't just about collecting certificates or joining forums. It's about being active and connected over time. Monthly challenge groups keep things interesting; these pop up in *Slack* channels or *Discord* communities and push you to build something new every few weeks. One month, you'll try creating a simple recommendation engine; the next, you might automate a personal workflow with no-code tools. Accountability circles, small groups of peers, help you stay committed to your goals and share wins or frustrations without judgment. Peer code reviews are another underrated tool: when someone else reviews your work (and vice versa), you spot blind spots more quickly and gain new ideas for handling edge cases or optimizing workflows.

Open-source AI projects are a goldmine for practical experience. Jumping into one might sound intimidating at first. There's jargon and unfamiliar code, but most projects welcome beginners who want to test features, write documentation, or improve usability. Even small contributions make your name visible in the field and provide you with stories to share in interviews or at networking events. Plus, collaborating across time zones with people from around the globe gives you a taste of how real-world remote teams function.

As you build skills through courses and grow your network in these communities, you'll naturally start spotting trends that matter for your work and filter out the hype that wastes your time. If you ever feel stuck or lost in the flood of new technology, remember that the combination of structured learning and an active support network will keep you moving forward.

Bringing it all together, certifications demonstrate your knowledge; courses provide structure (and sometimes a helpful nudge); communities keep you connected and inspired when motivation wanes. The next chapter will zoom out even further: we'll look at cutting-edge trends shaping AI's future so you can spot opportunities early and prepare for what's around the corner, not just what's hot right now.

Chapter Five

Bias, Fairness, and Responsible Use Are AI Ethics for Real-Life

Red Flags and Real-World Examples for Spotting Algorithmic Bias

Despite its reputation for objectivity, AI is only as fair and accurate as the data and design choices behind it. Algorithmic bias occurs when an AI system's decisions unfairly favor certain groups over others, which can have real-world consequences, including whether someone receives an interview, is treated fairly by public systems, or even if a social app accurately recognizes their face. Sometimes, bias is subtle, as seen in facial recognition systems that often fail to recognize people with darker skin tones. At other times, it's glaring, such as a hiring tool that was trained to favor men because it was primarily trained on male resumes. Bias isn't accidental; it emerges from biased historical data, developer assumptions, and embedded preferences. And when AI systems stumble, people can miss opportunities, be unfairly flagged, or be ignored by systems designed to help them.

AI bias is not just a technical concern; it has profound implications for people's lives. Consider the scenario of applying for a dream job, only to be filtered out not due to qualifications but because an AI system, like *Amazon's* hiring tool, has learned to reject resumes from women after being trained on a decade's worth of mostly male submissions. Or the use of facial recognition by police or at airports, which repeatedly misidentifies Black and Asian faces with dangerous consequences, such as wrongful stops or arrests. These are not isolated incidents: they stem from biased data and inadequate oversight, often with life-changing consequences.

Bias can enter unnoticed if no one scrutinizes how models are trained or what data they use. For example, a public agency might deploy facial recognition technology that performs well in tests but fails in real-world conditions, especially for underrepresented groups, because it overlooks gaps in its data. Similarly, a company might use an AI resume screener trained mostly on one demographic, resulting in predictable and unfair filtering. Without checks, minor issues become systemic injustices.

You don't need to be a tech expert to spot potential bias. Here's a practical checklist for evaluating any AI tool:

Checklist: Red Flags for Algorithmic Bias

- Does the tool show more errors for particular groups (e.g., women, people of color, non-native English speakers)?

- Is there any transparency regarding the training data sources, or is the information kept secret?

- Are outcomes reviewed regularly across different demographics?

- Has the system been tested outside of ideal, lab-like conditions?

- Is documentation available explaining how decisions are made and what happens if a problem arises?

- Is there an easy way for users to report unfair or incorrect outcomes?

- Are people from affected communities included in reviewing or improving the tool?

Consider these high-profile examples. The *Correctional Offender Management Profiling for Alternative Sanctions* (COMPAS) algorithm used in U.S. courts to assess re-offense risk was found to be twice as likely to wrongly mark Black defendants as high-risk compared to white defendants, a costly mistake, as it can lead to unjust incarceration. Resume screening systems, intended to streamline hiring, have perpetuated gender stereotypes and penalized "female-coded" language because of biased training. Predictive policing software, meant to reduce crime, often just directs more resources to already over-policed neighborhoods by learning from historic arrest data, reinforcing rather than correcting inequities.

If you suspect bias or something seems off, don't hesitate to speak up. You don't need advanced technical skills to make a difference. Document concrete examples of unfairness, collect supporting info and communicate clearly about the impact, not just the flaw. Your vigilance and questions can drive organizations to prevent mistakes and reputational damage; your role in this process is crucial.

"Hi [Manager/IT Lead],

I've noticed [describe the issue] with [AI tool]. For instance, [give a specific example: 'qualified candidates from diverse backgrounds are being flagged as unqualified more often']. This could reflect an underlying bias in training or usage. Can we review how it works and what safeguards ensure fairness? I want to discuss options for auditing or improvement."

When dealing with vendors, ask direct questions: "Can you share error rates broken down by demographic group?" "How diverse was your training data?" "What steps do you take if bias is reported?" Vague answers are another warning sign.

Bias doesn't need to be deliberate to have serious consequences. Raising these issues, even without a technical background, can drive organizations to prevent mistakes and reputational damage; your questions and vigilance do matter.

Practical Playbooks for Building Fairness into AI Decisions

Nobody wants to be on the wrong end of an unfair decision, especially when that decision comes from a system that's supposed to be "neutral." Fairness in AI isn't just a buzzword; it's a genuine necessity in any workplace or community that aims to treat people with respect and dignity. To make this happen, you don't need to be a coder or mathematician. There are clear playbooks and frameworks you can use, even if you don't touch code or data directly. The trick is understanding what "fair" looks like for your specific situation and then holding your tools and your team accountable to that standard.

Let's break down two of the big approaches to fairness you'll hear about: demographic parity and equal opportunity. Demographic parity refers to an AI system producing positive outcomes (such as job offers, college admissions, or loan approvals) at similar rates for different groups defined by factors like gender, race, or disability status. For example, if a university admissions algorithm is fair under demographic parity, students from all backgrounds would have similar acceptance rates, regardless of their group identity. But sometimes parity isn't enough. That's where equal opportunity (sometimes referred to as equalized odds) comes into play. This approach examines whether qualified individuals from each group have the same opportunity for a positive outcome, even if overall acceptance rates vary. For instance, in loan approvals, equal opportunity means applicants with the same financial profile get treated the same way, regardless of other traits.

Choosing which fairness framework to use depends on context. In university admissions, demographic parity may be desirable if you aim for a diverse student body. In lending, equal opportunity is often better since you want every qualified applicant to have the same shot at approval, not just to hit a quota. There's no one-size-fits-all. What matters is matching your fairness goal to your mission and values.

Getting fairness right starts with a clear plan. The first step is to pinpoint which groups require protection in your context. These might include age, ethnicity, gender identity, disability status, or language background, whatever's most relevant to your community or business. Next, set explicit fairness goals: "We want our AI hiring tool to offer interviews to candidates from all backgrounds at similar rates," or "Our credit scoring model needs to approve equally qualified applicants from every group." Then, measure current outcomes; don't just make an educated guess. Look at your data: Are there big differences in approval rates or error rates between groups? This is called measuring disparate impact, and it's how you spot whether your system is tilting the playing field.

Once you've identified gaps, it's time for action. You can start by reweighting the training data so that groups that were previously underrepresented or overlooked have a greater influence on the model's learning process. This might sound technical, but it boils down to telling your developers or vendors: "Let's make sure our training data reflects the diversity of people we serve." Another concrete move is asking for regular human review of cases that algorithms flag as unusual or high-risk. If an AI system suggests denying a loan to someone who looks qualified on paper, a real person should double-check before sending out the rejection email.

There are also practical tools and templates you can use right away. For example, HR departments can adopt a simple fairness policy: "We will audit all hiring algorithms each quarter for demographic parity across gender and ethnicity. Disparities above 5% will trigger a review." Sharing this policy with your team keeps everyone on the same page and signals that fairness is non-negotiable. You don't have to start from scratch, either; there are open-source toolkits designed for fairness auditing, such as IBM's AI Fairness 360 or Google's What-If Tool. These tools enable you to upload data and observe how different groups are treated by your models without requiring advanced technical skills.

To make these ideas concrete in the workplace, use a fairness impact assessment worksheet during the kickoff of any AI project. First, list the purpose of your system and its intended users. Then, jot down which groups could be harmed if things go wrong. Verify that these groups are well-represented in your training data or user testing pool. Finally, outline who will review outcomes for bias and how often those checks will happen.

Below is a quick template you can adapt:

Fairness Impact Assessment Worksheet

- **System purpose:** (e.g., Resume screening for entry-level roles)

- **Primary user groups:** (e.g., Recent graduates from diverse schools)

- **Protected groups considered:** (e.g., Gender identity, ethnicity, disability)

- **Fairness goal:** (e.g., Equal interview offer rates for all groups)

- **Data representation check:** Are all groups represented in the training and testing data?

- **Bias review interval:** How often will outcomes be checked? By whom?

- **Escalation plan:** Who gets notified if disparities are found? What steps follow?

Building fairness into AI isn't just about avoiding lawsuits or bad headlines; it's about treating people as equals and ensuring that your organization's technology reflects its values. Whether you're leading a team, evaluating a vendor product, or just raising concerns in a meeting, these playbooks and tools give you real ways to move fairness from a buzzword to daily practice.

What You Need to Know about Data Privacy in the Age of AI

AI has thrown a curveball into the world of data privacy. It's not just about securing passwords or protecting email addresses anymore. AI systems collect, analyze, and connect information in ways that can feel magical or scary, depending on your perspective. What makes AI different is its ability to piece together bits of harmless data and turn them into something surprisingly intimate. Your location pings, step counts, or even your online shopping habits might seem routine. Still, when an AI combines them, it can sometimes reveal things you never intended to share. For example, an app that tracks your mood and sleep may also flag behavioral shifts that indicate potential sensitive health conditions. AI's power stems from its ability to aggregate and infer, enabling it to spot patterns that humans might miss. This ability means a system could predict something as personal as pregnancy or depression risk, even if you never typed those words anywhere. All this adds up to new privacy risks: not only can AI "re-identify" individuals from supposedly anonymous data, but it can also guess details about your life you never volunteered.

This landscape places greater pressure on organizations and individuals to understand and respect privacy standards. The two major legal frameworks in this context are the European Union's *General Data Protection Regulation* (GDPR) and *California's Consumer Privacy Act* (CCPA). GDPR, which covers anyone dealing with EU residents' data, demands absolute transparency from anyone using AI to process personal information. One of its standout rules is the "right to explanation." Suppose an AI system makes an important decision about you, say, approving a loan or denying a job application. In that case, you can ask for details on how the system reached its verdict. Companies must be prepared to answer. The CCPA, meanwhile, gives California residents the right to know what data companies collect about them and to opt out of having that data sold or shared. Both laws are crystal clear on one point: people should have control over their personal information, and companies need to explain not just what they do with your data but also how their algorithms use it. Businesses that ignore these rules risk fines, lawsuits, public backlash, and loss of customer trust.

You don't have to be a privacy lawyer to spot trouble. When reviewing an AI project or tool, whether for work or personal use, it pays to ask a few pointed questions. Where is your data stored? Is it stored locally in your country or shipped overseas? Who else gets access, just the company behind the tool, or third parties as well? How long does the company hold onto your data after you stop using the service? Are there clear policies for deleting your information if you ask? If you're buying or recommending software at work, push vendors for detailed privacy disclosures. Do they encrypt sensitive info? How often do they conduct security audits? Can they provide a plain-vanilla summary of how user data is handled and protected?

Privacy Risk Assessment Checklist

- What types of personal data does the AI system collect?

- Where (physically) is the data stored?

- Who can access or receive this data?

- How long is user data retained after the service ends?

- Are there protocols in place for deleting data upon request?

- Does the vendor use encryption and regular security audits?

- Is there an accessible privacy policy written in clear language?

Most companies claim to care about privacy, but their actions often speak louder than the policies they link to at the bottom of their websites. Here's where you can take concrete steps, at work or even as a solo creator, to protect people's information. First up: minimize what you collect. If you don't need a piece of data to deliver your service, skip it. The less you hold onto, the less there is to lose if something goes sideways. For records you do keep, anonymize them wherever possible by stripping out names, addresses, phone numbers, or any other details that could be traced back to real people.

"Privacy by design" is more than a buzzword; it means baking privacy into every step of building or selecting an AI system. Ask yourself: if this tool got hacked tomorrow, what could someone steal? Default to strong protections; encrypt sensitive files and limit access to only those who need it. Make sure any consent language is easy for real humans to understand: "We use your activity data to personalize recommendations. You can change these settings anytime." Avoid legalese and jargon.

When interacting with users or customers, establish trust by being transparent about your privacy practices. A simple FAQ page can do wonders: What info do we collect? Why do we need it? How can you delete your account or review your data? Add contact information for privacy questions; sometimes, people want reassurance that a real person will listen if they raise a concern.

Staying on top of privacy isn't just about dodging regulators; it's about treating people's information with respect and building lasting relationships. Whether you're leading a project or just picking apps for your team, these habits keep you on solid ground in a world where data, once scattered and harmless, can quickly become deeply personal in the hands of a clever algorithm.

Human-in-the-Loop Is About Keeping People in Charge of AI Outcomes

Human-in-the-loop, or HITL for short, is a simple but powerful idea: people remain at the center of AI systems, guiding, correcting, and stepping in when algorithms fall short. Instead of letting the software run unchecked or hoping for perfect predictions, HITL means that someone, possibly you or a team member, has the power to pause, review, or override results. This isn't about distrusting technology. It's about recognizing that even the most sophisticated system can overlook context, miss subtle edge cases, or make mistakes. Your experience and judgment act as a safety net, catching what automation can't. In fields like medical image diagnostics, for instance, even when an algorithm highlights a suspicious spot on a scan, a skilled radiologist steps in to interpret the result. Machines can process vast volumes of images at lightning speed. Still, it's a human who puts those findings into perspective, spotting false positives or recognizing rare conditions that don't fit the usual patterns.

In any workflow where decisions carry real consequences, such as finance, healthcare, or content moderation, there are moments when human review isn't just helpful; it's non-negotiable. Take loan approvals as a practical example. An AI might process the bulk of applications, flagging apparent declines and approvals. However, for borderline cases or applicants with unusual financial histories, a loan officer reviews the file before making any final decision. That way, someone can spot if a hardworking entrepreneur gets penalized just because their income doesn't fit a traditional mold. In online platforms where AI filters handle content moderation, users need an appeals process. If someone's post is taken down by mistake, a human moderator may review it and decide if it should be restored. These checkpoints ensure fairness and prevent tech from shutting people out unfairly.

Striking a good balance between automation and human oversight isn't always obvious. There's real power in letting machines handle routine, repetitive tasks. They're fast, tireless, and don't get bored. But when stakes rise, automation without oversight is risky. One way to sort this out is by using a risk matrix, a framework that helps you decide when to let AI take the wheel and when to require human sign-off. Low-risk tasks (like sorting spam emails) can be fully automated. Medium-risk actions (like customer service replies) might call for spot checks or random audits. High-risk scenarios (like medical diagnoses or large financial transactions) need explicit human approval every time. It's not just about covering your bases; it's about protecting people from costly mistakes that can occur when systems act on incorrect data or encounter scenarios they've never experienced before. Anyone who has seen an automated system lock out hundreds of users due to a false positive knows how expensive and embarrassing unchecked automation can be.

Embedding HITL into your organization doesn't happen by accident; it takes practical steps. Start with clear standard operating procedures (SOPs) that outline exactly when and how humans are involved in the process. For instance, your SOP might require that a supervisor review any flagged edge case in a loan process within 24 hours. Or your content team might have rules for when moderators must manually check posts that trigger specific sensitive keywords. These procedures keep teams aligned and ensure no one skips steps under pressure.

Training is another cornerstone for effective HITL systems. It's not enough to hand out a checklist and hope for the best. Frontline staff, whether they're reviewing flagged transactions or moderating challenging posts, need tailored modules that introduce them to how the AI tool works, its limitations, and how to escalate when something feels off. It's also smart to run through real-world scenarios during training: "Here's what happens when the AI flags something unexpected; here's how you respond." This builds confidence and keeps everyone alert to subtle glitches.

Good documentation brings everything together. Every critical decision point where someone intervenes to review or override should be documented, not just for compliance reasons but also so teams can identify patterns over time and refine the system for better results. Templates can be helpful here: set up simple forms where reviewers note what prompted them to step in, the judgment they used, and the outcome they chose. Over time, this archive becomes invaluable; if similar issues arise again, you won't have to reinvent the wheel every time.

The more transparent your organization is about these decision points, the easier it becomes for everyone, from new hires to C-suite leaders, to understand where humans add value and where automation is sufficient. When people know precisely where their judgment is needed (and where it's not), teams move faster, make fewer errors, and build more trust with customers who want to know real people still care about their experience. In workplaces where the pressure to automate everything is high, keeping humans at the heart of AI-driven workflows isn't old-fashioned; it's innovative risk management and the best way to ensure technology serves people, not the other way around.

Human-in-the-loop thinking isn't just about plugging gaps. It's about building systems where people and algorithms work together seamlessly. This balance enables you to scale up efficiency without compromising ethics or accountability. It provides employees with meaningful roles in guiding technology, rather than replacing them overnight, which also helps boost morale and foster buy-in for new tools. As AI becomes increasingly integrated into everyday work, organizations that prioritize human oversight position themselves for resilience and maintain their credibility when things become complicated or messy.

Communicating Risks and Limitations to Stakeholders Ensures Transparent AI

Trust in AI doesn't happen because someone says, "Trust me." It's built on candor and open communication about what a system can do, where it might fall short, and what those boundaries mean for the people using it. Suppose you've ever watched a product fail in public because its creators hyped up magic but delivered headaches. In that case, you know how quickly trust can unravel. I still remember the uproar when a major tech company demoed a "fully autonomous" assistant, only for users to discover it was prone to hilarious, sometimes embarrassing mistakes. Customers felt misled, and the company spent months repairing relationships and rewriting its marketing pitch from scratch. That backlash wasn't just about bugs; it was about the gap between promise and reality. When teams speak plainly about risks and limitations, even tough news is easier to accept. People want to know what they're dealing with, not a fairy tale.

If you're responsible for rolling out an AI system or explaining its impact, you need more than good intentions; you need a toolkit for clear, audience-appropriate communication. When I prep for board meetings or executive updates, I start with a one-page summary: "Here's what the AI system is designed to do. Here's where it works brilliantly. And here's where it struggles or needs human backup. These are the risks we're tracking, and these are the plans for review." No jargon, no smoke and mirrors, just direct answers to the questions leaders care about. If you're writing customer-facing info, an FAQ is your best friend. Start with the basics: "How does AI affect my service?" "Can I opt out?" "What happens if the AI makes a mistake?" Add answers that are honest, concise, and genuinely helpful. For example: "Our AI suggests responses to help our team reply faster, but real people always review before sending." This builds confidence and demonstrates your commitment to transparency.

The real challenge arises when discussing limits, the boundaries where your AI might falter, or when you can't guarantee perfect results. People respect honesty far more than empty promises. When drafting risk disclosures for marketing or press materials, I focus on specifics: "This tool predicts trends based on historical data and may underperform during unprecedented events." For product managers or client-facing teams, scenario planning is a lifesaver. Map out "what-ifs," such as "What if our AI fails to detect an issue?" and write responses in advance. You might say, "While our model catches 95% of errors, we've put manual spot checks in place for the rare situations where it falls short." This kind of prep not only boosts your team's credibility but also reassures users that you're not hiding behind the tech.

Embedding transparency into an organization requires more than a single memo or policy update; it's about cultivating a culture of transparency. Teams that value openness build channels for ongoing conversation: monthly "Ask Me Anything" sessions with engineers or data scientists can demystify technical choices and invite tough questions from users or colleagues. Internal workshops help non-technical staff recognize both the power and pitfalls of AI so unexpected results catch them off guard. I've seen companies run short exercises where employees try to "break" an AI tool, surfacing blind spots and sparking honest talks about what happens when things go sideways.

Routine transparency reporting also sets expectations. Instead of waiting for complaints to surface, send regular updates that cover key performance stats, error rates, and known limitations. A simple quarterly email could include: "Here's how our AI performed this month; here's what we're improving; here's who to contact if you spot a problem." Templates make it easy for teams to maintain consistent communication without having to reinvent the wheel every time.

Bringing transparency into every step of your AI process doesn't just protect your brand; it empowers everyone involved (users, staff, and even executives) to engage with technology on their terms. When people are kept in the loop, especially when things don't go as planned, they're more likely to support your efforts, offer constructive feedback, and help shape solutions that work in real life. Clear communication closes the distance between hype and reality, making sure everyone knows where things stand.

As this chapter concludes, remember that ethical AI is more than a checklist; it's a practice that begins with honest words and evolves through shared responsibility. As you move forward, keep transparency at the center of your work. It will serve you well in every conversation about technology's role in your organization and beyond. Next up: see how practical playbooks turn good intentions into tangible results, helping teams use AI not just responsibly but fearlessly.

Chapter Six

Master AI Communication by Explaining Complex Tech in Plain English

Crafting the Perfect AI Elevator Pitch for Your Team

Have you ever tried explaining AI at work and watched your audience's eyes glaze over? Maybe you started with "machine learning models" or "algorithmic optimization," only to notice people checking their phones. Almost everyone has felt the pressure of making a new idea take hold, especially when it comes to technology that feels mysterious or intimidating. The truth is, you don't need a computer science degree or slick presentation skills to get folks on board with AI. What you do need is a clear and concise pitch that speaks directly to what matters most to your audience. Think of it like a trailer for a blockbuster: short, punchy, and focused on what people care about. If you can break down AI in 30 seconds, you're light-years ahead of most.

Let's start with why that matters. In the real world, attention spans are short. You might have a ride in the elevator or the time it takes to grab a coffee to capture interest. The key isn't to sound impressive but to make AI feel relevant and valuable. That means ditching the technical jargon and highlighting practical outcomes. For example, if your team dreads data entry, you might say, "This tool automates data entry, giving us back ten hours every week." Or, if someone is always double-checking spreadsheets for mistakes, try: "It flags potential errors so we catch them before they snowball." These are clear wins that resonate with busy professionals, relieving them of tedious tasks and giving them hope for a more efficient work process.

When crafting your elevator pitch, it's helpful to follow a simple structure. This keeps you on track and ensures you cover all the bases, including problem, solution, and benefit, without rambling.

My go-to formula looks like this:

"Our team spends [X] hours on [Y task]. AI helps by [Z], letting us focus on [W]."

It's fill-in-the-blank simplicity with maximum punch.

Here's how it plays out:

"We lose six hours every week sorting through support tickets. AI sorts and prioritizes them for us so we can focus on helping customers faster."

Notice how it centers on a pain point that everyone recognizes, then focuses on the value delivered.

Tailoring your pitch to your audience is where the magic happens. The concerns of someone in HR aren't the same as those in sales or operations. For sales teams, consider using "AI recommends the best leads based on recent behavior, so we close more deals faster. If you're talking to HR: "This tool screens resumes for us, helping spot top talent quickly and reducing unconscious bias." Operations? Try: "AI forecasts inventory needs so we never run out or overstock, saving us money and hassle." These aren't generic one-size-fits-all lines; they show you what each department needs.

There are situations where you'll face skeptics or challenging questions, maybe from a colleague worried about job security or a leader who's been burned by overhyped tech before. Here's how to handle those moments with empathy and confidence. If someone asks, "Will this replace my job?" don't dodge or sugarcoat it. Say something like, "No, it's designed to take tedious work off your plate so you can focus on projects that need your creativity and judgment." This shifts the narrative from fear of replacement to opportunity for growth and impact.

When leadership wants proof, "How do we know it'll work?" lean on pilots or early results. Instead of promising the moon, point to specific outcomes: "During our two-week pilot, AI reduced manual errors by 30% and saved us five hours per person." Keep it grounded in facts and real-world wins. If you don't have results yet, set clear expectations: "We'll start small, measure the impact, and decide together if it's worth scaling."

Build Your Own Pitch

Take a minute and jot down answers to these prompts:

- What's one repetitive or pain-in-the-neck task your team deals with?

- How could AI help (automate, flag errors, prioritize, etc.)?

- What would your team do with the time or energy saved?

Now plug your answers into this template:

"Our team spends [X] hours on [Y]. AI helps by [Z], letting us focus on [W]."

Example:

"Our marketing crew spends four hours weekly tagging social posts. AI auto-tags content based on trends so we can focus on creative campaigns."

Run this by a friend or coworker. Notice what grabs their attention—and what doesn't land. Adjust your language until it sounds natural and specific.

If you're pitching to a group with mixed roles or backgrounds, choose examples that bridge their interests. For instance: "AI automates routine paperwork across departments, which means fewer bottlenecks for everyone and more time for strategic projects." This approach brings people together around shared pain points and collective wins.

The art of a strong elevator pitch isn't about dazzling anyone with technical detail; it's about showing that you understand their needs and have a practical solution that makes life easier. When you speak their language and answer their real concerns, whether it's about time, error reduction, fairness, or job satisfaction, you turn AI from an intimidating buzzword into an everyday advantage they actually want.

Consider Sources such as:

- 11 great elevator pitch examples & how to make yours

- How to explain AI to non-technical stakeholders

How to Make Data and Models Relatable by Storytelling with AI

People connect with stories, not dashboards. Suppose you've ever watched colleagues tune out during a technical meeting. In that case, you know the pain of seeing good data go ignored because it feels too abstract. The trick isn't just to present numbers or model outputs; it's to wrap them in a narrative that resonates with your audience. This means taking the raw output from an AI system and transforming it into something that feels real and useful to people who care about outcomes rather than algorithms. For teams trying to get AI projects off the ground, it's often the story, not the spreadsheet, that gets buy-in. When you can say, "Our new chatbot solved 200 real customer problems last month, and almost everybody gave it a thumbs up," you're speaking their language. Suddenly, AI isn't some mysterious code box; it's the reason a customer left happy instead of frustrated.

Compelling data storytelling starts by aligning AI results with what matters to your business or community. If you're trying to show the value of an AI tool, focus on wins people recognize. Your IT team used to spend hours resetting forgotten passwords; now, with an AI-driven self-service portal, those requests have dropped by 80%. Frame it as a clear before-and-after: "Before, users waited overnight for help; now, they must reset their password instantly." This way of structuring stories — challenge, solution, result — makes it easy to see how AI has a direct impact. The best stories use specific numbers while maintaining a focus on people. "Support wait times used to stretch for days; now, thanks to an automated triage system, most problems get solved in under two hours." These details transform a dry stat into a win that everyone can picture and celebrate.

Putting a face on AI outcomes is where stories come alive. Data alone is forgettable. Pairing it with personal or team experiences makes AI feel like a force for good in real life. Perhaps a sales representative named Jordan utilized an AI-powered lead-scoring tool and finally closed a deal that had been stuck for months. Let Jordan tell that story in their own words. Or share a customer's feedback after using your new AI helpdesk: "After struggling with my order for a week, the virtual assistant walked me through every step, problem solved!" These personal vignettes provide context to the numbers, making your case relatable to everyone, from frontline employees to top executives.

Building your own compelling AI stories doesn't require years of writing experience, just a repeatable structure and a few prompts to guide you. The classic arc works every time:

Problem → Struggle → AI Solution → Win.

Start by describing the issue in terms everyone understands. Spell out what made it painful or annoying: long waits, repetitive work, costly mistakes. Next, highlight the efforts or frustrations that occurred before AI stepped in. Then, introduce your solution: "We rolled out an AI scheduling bot that syncs everyone's calendars in seconds." Finally, land on the win. Quantify it if you can, hours saved, complaints reduced, revenue gained, or describe the change in morale or workflow.

To get your creative juices flowing, lean on prompts like:

- Who benefited most from this change?

- What surprised us when we launched?

- What did people do differently after AI entered the picture?

- Was there resistance at first?

- How did that shift once people started seeing results?

- Did any unexpected "aha!" moments crop up?

These questions help you move beyond surface-level wins and get at what changed for people in their daily routines, their mood at work, or even their willingness to try new things.

Emotion matters, even with technical stories. Don't be afraid to mention relief, excitement, or even initial skepticism. "At first, our support staff worried the bot would do more work for them, but after two weeks of fewer angry emails and simpler escalations, they were asking when we'd automate other drudge tasks." This kind of emotional honesty makes stories stick and helps others see themselves in your narrative.

Story Arc Toolkit

Try this fill-in-the-blank exercise to build your own story:

- **Problem:** What was the pain point? Who felt it?

- **Struggle:** How did people cope or think before?

- **AI Solution:** What tech did you introduce? How did it work?

- **Win:** What changed—metrics, morale, customer happiness?

Example:

Problem: Our HR team was buried under hundreds of resumes every week.

Struggle: The process was slow, and we missed great candidates who didn't use the right buzzwords.

AI Solution: We introduced an AI screening tool that highlights top matches based on fundamental skills and experience rather than just keywords.

Win: Now our HR staff spends less time sifting and more time interviewing strong talent; one manager said it finally "put the human back in HR."

Keep stories short but vivid. Use names and quotes if possible. Even one strong example can inspire others across your company to view AI not as cold automation but as something that genuinely makes work and life a little better.

How to Run Effective AI Demos From Setup to Stakeholder Buy-In

Prepping for an AI demo can feel daunting, especially when many in your audience are skeptical. The goal isn't to impress with complexity but to make the technology feel simple, practical, and relevant. Always start by considering your audience. What do they care about? Tailor your demo data to reflect their world. If you're presenting to HR, use samples that mimic standard hiring spreadsheets. For customer service folks, bring sample support tickets. The closer your demo mirrors their real work, the more engaged they'll be.

Keep the demo brief, ideally lasting less than ten minutes, and aim for a duration of around five minutes. Audiences tend to lose focus quickly, especially with unfamiliar technology, so rehearse your walkthrough multiple times to ensure it remains clear and concise. Start solo, then ask a candid colleague to point out confusing parts or rough transitions. This helps polish your delivery and ensures every step directly relates to a real benefit for your audience.

When it comes to demo data, steer clear of unrealistic or generic examples, such as "John Doe" or "Test Account." Use actual (but anonymized) data that matches realistic scenarios. If that's not possible, base your dummy data on real patterns from your audience's workflow. This authenticity makes a big difference in how believable and valuable your demo feels. Always double-check data for accuracy and privacy; use anonymized names or placeholder data if necessary.

---

cold automation but as something that genuinely makes work and life a little better.

How to Run Effective AI Demos From Setup to Stakeholder Buy-In

Prepping for an AI demo can feel daunting, especially when many in your audience are skeptical. The goal isn't to impress with complexity but to make the technology feel simple, practical, and relevant. Always start by considering your audience. What do they care about? Tailor your demo data to reflect their world. If you're presenting to HR, use samples that mimic standard hiring spreadsheets. For customer service folks, bring sample support tickets. The closer your demo mirrors their real work, the more engaged they'll be.

Keep the demo brief, ideally lasting less than ten minutes, and aim for a duration of around five minutes. Audiences tend to lose focus quickly, especially with unfamiliar technology, so rehearse your walkthrough multiple times to ensure it remains clear and concise. Start solo, then ask a candid colleague to point out confusing parts or rough transitions. This helps polish your delivery and ensures every step directly relates to a real benefit for your audience.

When it comes to demo data, steer clear of unrealistic or generic examples, such as "John Doe" or "Test Account." Use actual (but anonymized) data that matches realistic scenarios. If that's not possible, base your dummy data on real patterns from your audience's workflow. This authenticity makes a big difference in how believable and valuable your demo feels. Always double-check data for accuracy and privacy; use anonymized names or placeholder data if necessary.

Avoid the pitfall of trying to show too much; don't overwhelm your viewers with dashboards and every feature. "Death by dashboard" is a real phenomenon; displaying countless menus and charts only confuses and bores people. Instead, focus on one or two crucial features. Narrate each step simply: "See this alert? That's how AI lets us know which customers are at risk of leaving." Or "This feature highlights compliance risks before they turn expensive." Clearly connect features to real-world workplace problems.

Technical issues are inevitable, so always bring a backup: screenshots, screen recordings, or a slide deck summarizing key points. If your live demo fails, switch to the backup and keep rolling. This saves face and maintains credibility.

The most memorable demos clearly tie back to real business value. As you present each feature, connect it directly to important outcomes: "This dashboard isn't just visually appealing. It pinpoints exactly where we're losing customers." Or "This feature spots compliance issues before they become costly." Don't just showcase what AI does. Show why it matters and which pain points it solves. The goal is for your audience to see themselves and their challenges reflected in your presentation.

When you finish the demo, don't just close your laptop and leave. Use the opportunity to gather feedback that can improve your presentation and the potential rollout. Ask questions like:

- "What did you find most useful?"

- "Did anything seem confusing?"

- "Where might this tool help your daily work?"

- "Any concerns about how this fits with your workflow?"

Listen carefully and take notes. These insights are invaluable for refining your approach. After the meeting, follow up with a brief email: "Thanks for your time today! Based on your feedback, I'd love to explore testing this in [your department]. Are there colleagues who should see a shorter demo?" This shows you're collaborative and focused on real needs rather than just pitching tech.

Encourage questions and objections instead of ignoring them. Some will ask about data privacy; others may worry about workflow disruption. Acknowledge these honestly and suggest a small, safe next step, like, "We can start with a limited pilot using only anonymized data," or "Let's run a test with one team first to work out any kinks." This builds absolute trust and buy-in, far more than a flashy feature ever could.

Use plain, direct language grounded in your audience's reality. Skip technical jargon unless specifically asked. Even if you know the inner workings, your job is to make others feel empowered, not overwhelmed. When faced with hard-to-explain features, use helpful analogies: "Think of this alert system like your car's check engine light; it lets you know there's a problem and where to look but doesn't fix it for you."

Great demos balance clarity with excitement. Leave your audience curious and eager to learn more while also confident that AI can alleviate pain points and make their day-to-day work smoother and less stressful.

A Tactical Guide to Debunking AI Myths in the Workplace

Misinformation about artificial intelligence spreads rapidly, especially in workplaces, think break rooms, meetings, and team updates. Many people have picked up alarming stories or half-truths, unknowingly carrying around a suitcase full of AI myths. If you've ever heard phrases like "AI will replace all our jobs" or "AI is totally objective," you know it's hard not to roll your eyes. Let's break down the ten most persistent AI myths in today's offices, paired with clear, evidence-based answers you can use.

First up, the favorite: "AI will replace everyone's job." This myth persists, but the reality is that AI primarily automates repetitive tasks, such as data entry, invoice sorting, or flagging duplicates. It frees people up for creative problem-solving that needs a human touch. Another myth: "AI is always unbiased." That is not true; AI learns from data, which can reflect human bias. If bias is in the data, the AI repeats those patterns. Then there's "AI can learn anything by itself," but AI needs lots of labeled data and regular human updates to improve. Similarly, the statement "AI understands context just like a person" is off base; most AI lacks common sense, relying solely on patterns.

Other myths include: "Once you set up AI, it runs itself." In fact, AI requires ongoing monitoring, retraining, and adjusting. "Only big tech companies can afford AI" isn't true either. Numerous affordable and accessible tools are available for small businesses and nonprofits. There's also the idea that "AI will solve every business problem." Not so; sometimes simpler solutions are better, and AI can't fix broken processes or poor management. Some believe, "If it's labeled 'AI-powered,' it's advanced." Often, that's just marketing hype; some "AI" tools are basic automation. Another: "AI is 100% accurate." In reality, every AI model makes mistakes, especially with messy or unfamiliar data. Finally: "Using AI means we don't need humans." The best results are achieved by combining people and AI. Humans set the goals, review results, and make decisions.

To debunk these myths effectively, use empathy, not just facts. People naturally worry about job loss, fairness, or privacy. If someone asks, "Will this AI tool automate my job?" respond with understanding: "That's a common fear; lots of headlines talk about job loss. But so far, automation usually just means less repetitive work and more meaningful, customer-facing projects." If someone says, "AI is colorblind and doesn't discriminate," gently clarify: "That would be wonderful! But studies show

AI can pick up on bias in training data, so we need to monitor how it's built and used."

Keep conversations casual and respectful. When someone repeats a myth, like "AI will make all decisions for us," instead of correcting them bluntly, try: "Actually, what I've seen is most systems need people to set rules and check results. AI does heavy lifting on data, but people are still in charge." Or if someone claims "AI never makes mistakes," say, "I wish! Every system gets things wrong, especially with unfamiliar info, so we always double-check." Quick, helpful, myth-busting scripts keep you friendly, not dismissive.

To truly shift mindsets within your team, encourage critical thinking rather than passively accepting AI hype. Suggest a monthly "AI Mythbusters" lunch session where anyone can bring up AI claims they've heard or articles they've read, then research what's real as a group. Share credible news from reputable sources, such as Wired or MIT Technology Review, rather than relying on hype-driven headlines and social media. Create an "AI claims" channel on Slack or Teams where anyone can post rumors to fact-check—no judgment, just honest curiosity.

You can also make learning interactive: run quizzes or polls about common misconceptions ("True or false: AI can make ethical decisions on its own?"). This encourages people to engage in conversation and reflection rather than simply absorbing information. When someone suggests a new tool or automation upgrade, encourage at least one critical question: "How do we know this works for our customers?" or "What happens if the data changes?" Over time, this fosters a culture of healthy skepticism, where people ask for evidence and consider risks rather than blindly accepting claims.

The more comfortable your group is with questioning AI stories together, the less myths can spread fear or friction. It also helps everyone feel included in changes rather than being at the mercy of others. You don't have to be an expert to spot hype, stay curious, ask questions, and keep the conversation open. This approach helps your team grow and allows them to enjoy debunking wild rumors along the way.

Methods for Visual Thinking for AI Using Infographics, Diagrams, and Whiteboard Sessions

Visuals significantly enhance information absorption and retention. Humans remember approximately 80% of what they see and do, compared to just 20% of what they read and 10% of what they hear. With AI's abstract nature, clear visuals make complex concepts tangible, bridging the gap between confusion and real understanding. That's why the best AI communicators rely on infographics, diagrams, and even quick whiteboard sketches to break down processes and clarify how data transforms into business outcomes.

For example, to illustrate how information transforms in an AI workflow, use an "AI pipeline" infographic. Imagine a simple left-to-right flow: raw data (like a pile of emails or receipts) on the left, arrows to a processing box (the AI analyzing patterns), and outcomes (such as a prioritized customer list or fraud alert) on the right. Use easily recognizable icons and high-contrast colors for each step (e.g., blue for data, green for processing, and yellow for results), maintaining a linear and uncluttered layout. Another effective format is the "decision flow" diagram. If you're implementing an AI-powered approval system, begin with the input (a loan application), map the lines to decision points (such as credit score and fraud check), and then branch out to outcomes (approved, needs review, or declined). These diagrams not only convey information but also make the logic visible.

Live whiteboard sessions are valuable for collaborating with teams. Real-time drawing brings concepts to life and invites engagement. Organize your space with markers and sticky notes. For something like chatbot logic, draw each key step as a box, such as "Greet Customer," "Ask for Order Number," and "Provide Status," and connect them with arrows to show possible paths. Get everyone to contribute with sticky notes to fill in the transitions, or ask the group to map out the data flow from "Input Data" to "Desired Outcome." This hands-on, visual approach demystifies abstract AI steps, encouraging clearer group understanding.

Accessibility is crucial when designing diagrams or infographics. Not everyone processes visual information in the same way. Choose high-contrast color schemes (such as dark blue on white or black on yellow) and avoid red-green color pairings to accommodate color blindness. Use large, readable sans-serif fonts, such as Arial or Verdana, and annotate each part; never assume that icons are self-explanatory. Always include descriptive alt text for visuals shared online, so that screen readers can interpret them accurately.

For an AI sales forecast infographic, stick to three clear blocks: "Sales Data Collected" (spreadsheet icon), arrow to "AI Model Predicts Trends" (gear icon), and arrow to "Forecast Used in Planning" (calendar icon). Clearly label each section, "Data In," "AI Analysis," and "Actionable Forecast." Add concise callouts for details like "Key variables: product type, seasonality," using text boxes. Choose an accessible color palette (such as navy blue, teal, or orange), and test your design in grayscale to ensure clarity.

You don't need complex software to create compelling visuals. *Canva* is simple for creating infographics with drag-and-drop shapes and icons. *Miro* offers digital whiteboards for remote teams, letting everyone co-create and save ideas. *Lucidchart* works well for more technical, structured diagrams. All these tools provide templates to speed up your workflow and eliminate the need to start from scratch.

Test your visuals with a fresh set of eyes. Share diagrams with a colleague who is uninvolved in their creation and ask for honest feedback on clarity and potential areas of confusion. This helps identify what requires further explanation or clearer labeling. Use their questions to refine your visuals with extra arrows, notes, or context as needed.

Best practices for AI visuals: maintain strong color contrast, use readable fonts at all sizes, label everything directly, add alt-text for all online visuals, keep each graphic focused on one main idea, avoid clutter by spacing out elements, and routinely check accessibility using tools or by collaborating with colleagues who use screen readers.

In summary, visual thinking is essential for communicating complex AI concepts. Effective diagrams don't just explain; they invite collaboration and spark new ideas. As you continue advancing with AI, keep these visual strategies central to your toolkit. Up next: practical playbooks for integrating AI into real-life workflows so you can put these visual skills to work where they matter most.

Chapter Seven

Here's What's Next in AI for Trends and Hype

Generative AI Unpacked Using Text, Images, Video, and Audio

Have you ever come across a captivating image or a catchy tune online and wondered about its origin, whether it was created by humans or AI? This uncertainty is becoming increasingly common. Generative AI is at the forefront of this shift, setting itself apart from earlier AI systems that primarily processed or organized existing data. Instead, generative AI is a force of creation, producing original content in various forms, including text, images, music, and more. It's not just about categorizing your vacation photos anymore; it's about crafting new poems, art, and even business reports, sparking a wave of inspiration and creativity.

Generative AI relies on neural networks and digital systems trained on massive datasets. While older AI systems might identify a dog in a photo or filter out spam, generative models can author complete articles, create digital art, compose music, and edit videos. Text generators like

ChatGPT and *Jasper* don't just autocomplete—they write entire blog posts, craft clever headlines, and help brainstorm difficult emails. Learning from billions of words, they can replicate tone, humor, and style, making them a favorite for marketers drafting newsletters or sales pitches.

Image generation goes a step further. Tools like *DALL-E* and *Midjourney* let you input a simple description, such as "a panda skateboarding in Times Square at sunset," and quickly generate detailed, unique images. Artists use these for inspiration and rapid mock-ups, while agencies create quick, compelling ads. Even those with no design skills can now visualize concepts without needing an illustrator. Video synthesis has advanced as well. *Runway ML*, for instance, creates short video clips from text or sketches, letting content creators and small businesses produce compelling videos with minimal resources.

Audio and music generation is also evolving. Apps like *Aiva* produce original scores for podcasts or videos. At the same time, *ElevenLabs* generates realistic synthetic voiceovers in various languages and tones. These AI voices can convey emotion and nuance, enabling brands to create multilingual content quickly and affordably.

Industries are adapting fast to the potential of generative AI. Consultants use generative AI to draft reports, thereby freeing up time for more in-depth analysis. Ad agencies generate visuals at scale to test what captures attention before ramping up production. Video editors localize content worldwide using synthetic voiceovers, avoiding the complexity of hiring global talent. In education, teachers design lessons with custom AI illustrations and tailored practice questions for students. In healthcare, generative AI is used to analyze medical images and assist in diagnosis. These are just a few examples of how generative AI can be a powerful tool for businesses, saving time and resources while enhancing productivity.

Yet, these advances also bring challenges. Deepfakes, or realistic fake videos created with generative AI, have fooled viewers and can spread misinformation rapidly [*Ethical Considerations & Copyright - Norwalk*

AI in Education]. Copyright concerns are growing: if an AI is trained on countless online artworks or songs, who owns what it produces? The US Copyright Office is starting to address this, but clear guidance is scarce.

Bias is another issue that must be considered when utilizing generative AI. Suppose a model's training data is biased towards certain cultures or internet stereotypes. In that case, its output can reflect and amplify those biases. This can lead to the perpetuation of stereotypes or the marginalization of certain groups. As generative AI handles more creative tasks, some worry about the future of human jobs, particularly entry-level roles in writing or design. Some view these tools as creative partners, while others see them as possible replacements.

To harness the power of generative AI effectively, it's best to start small. Experiment with an AI tool to draft a blog post or a social media caption. Continually review and edit before publishing, noting what needs improvement and what accelerates your work. This practical approach to experimentation will empower you to leverage the potential of generative AI in your work.

Your First Generative AI Experiment

Pick a generative AI tool, such as *ChatGPT* for text, *DALL-E* for images, or *ElevenLabs* for audio, and set a 30-minute timer. Identify a practical task (a blog intro, product image, or podcast script), then use the tool to generate it. Note what surprised you (whether good or bad), and edit until you're satisfied with the final result. Reflect on how it saved time, sparked ideas, or needed your judgment to improve it. This exercise helps you safely test the real value of generative AI in your workflow.

When exploring these innovative tools, it's crucial to approach them with both curiosity and caution. By remaining vigilant about the opportunities and the risks, you can ensure that generative AI remains a tool under your control, not one that controls you. This responsible approach to *using generative AI will help you navigate its potential pitfalls and maximize its benefits.*

Large Language Models (LLMs), What They Are and How to Use Them

Trying to wrap your head around large language models can feel like chasing a moving target. One week, you hear about "GPT-4" in a team meeting. The following week, you see *Claude* or some new chatbot trending online. At their core, LLMs are like the "supercharged autocomplete" you wish your phone had. Imagine texting, but the predictive suggestions don't just finish your word. They spin out paragraphs, summarize long emails, or even pen a first draft of your monthly report. Unlike older AI systems, which primarily identified keywords or followed rigid scripts, LLMs can understand nuance, context, and the rhythm of everyday conversation. They've been trained on a mind-boggling amount of information and think everything from news articles to classic novels and tech blogs. This vast exposure enables them to predict not only what comes next in a sentence but also how to match the tone, mimic style, or sound helpful and professional.

The structure behind these models is complex, but you don't need to know all the technical details to get value from them. What sets LLMs apart is their ability to comprehend and interpret meaning across extensive text passages. Instead of analyzing just a few words at a time, they draw on entire paragraphs or conversations to gain a deeper understanding. This makes them feel eerily fluent. When you prompt *GPT-4* with "Summarize this contract," it not only selects key terms but also identifies clauses, flags risks, and suggests clarifying questions. It's as if you have an assistant who reads everything ever written and can instantly adapt to your requests.

In professional life, LLMs excel in areas where language becomes complex or repetitive, such as legal and medical fields. Drafting emails suddenly feels less like a chore. You pop in bullet points, and out comes a polished note ready for your client or boss. Proposals and reports are no longer blank-page nightmares; you feed in your main ideas and let the model build a structure, add detail, and even suggest improvements. For anyone drowning in documents, LLMs are lifesavers. Drop in a 20-page whitepaper or lengthy PDF, ask for a summary, and get the key points back in plain English—ideal for prepping meetings or sharing updates with your team.

Brainstorming also gets a boost. If you're stuck searching for creative taglines or campaign concepts, LLMs can offer dozens of ideas in seconds. Marketing teams utilize them to craft social media posts, generate ad copy variations and draft press releases. In design or product development, you can use LLMs to draft feature lists, customer FAQs, or onboarding scripts that sound engaging and clear.

Getting the best out of LLMs takes some practice. This is where "prompt engineering" comes in. The more specific and clear your request, the better the output. Instead of saying, "Write an email," try, "Draft a friendly follow-up email thanking a client for their feedback and proposing a quick call next week." Templates save time: for customer service replies, start with "Respond to this customer complaint about late delivery, apologize, and offer a discount." If the first response falls short, consider tweaking your prompt, adding context, or clarifying the desired tone. You might try "Make it more formal" or "Shorten to three sentences." This back-and-forth refines results until they fit your needs perfectly.

When creating an FAQ page for your website using an LLM, start by feeding it actual questions from real customers. Ask for concise answers in your brand voice. Review the first batch. If some answers are too long or off-topic, tell the model to "focus on clarity" or "avoid jargon." Each time you edit your prompt and regenerate answers, you'll notice the output getting closer to what works for you and your audience.

Still, it's crucial not to treat LLMs as infallible experts. Sometimes, they "hallucinate," confidently making up facts or quoting sources that don't exist. This isn't intentional; it's a side effect of how they predict text based on patterns from training data. Always double-check facts before forwarding anything important or publishing content externally. Another pitfall is bias. If their training data leaned toward certain viewpoints or stereotypes, those can slip into generated text without warning. It's wise to review outputs with a critical eye, especially when handling sensitive topics or company messaging.

Privacy is another area to watch closely. Never paste confidential client data or sensitive business information into public large language models (LLMs) unless you're sure about how the data is handled. Some platforms store prompts for future training or quality improvement purposes. Therefore, check the privacy policies before sharing anything that shouldn't leave your organization.

To stay responsible with these tools, treat LLM outputs as drafts, not finished work; review all claims; keep sensitive details out of prompts; and always ask yourself if an answer sounds too good (or too weird) to be true. As LLMs continue to evolve, those who use them thoughtfully, striking a balance between speed and judgment, will stay ahead while avoiding embarrassing mistakes or ethical missteps.

If you're new to all this, consider starting with a safe internal project, summarizing internal newsletters, or brainstorming ideas for your next team event. As you become accustomed to guiding the model's responses through careful prompting and review, you'll find more ways to incorporate this technology into everyday work without letting it run amok. The real magic isn't just in what these models can say; it's in what they can do. It's in how well they help you communicate faster and with less stress than ever before.

Edge AI in Smarter Devices and Smarter Homes

Your thermostat is learning your habits and adjusting itself before you even notice. Now, apply this across your entire house, including lights, security cameras, and voice assistants, all reacting in real-time, with no cloud lag and no waiting. That's the promise of Edge AI. Instead of offloading tasks to distant, powerful servers, Edge AI brings intelligence directly onto devices, like moving the brain into your own home. The result: decisions are faster, more private, and can often happen even without internet access.

Edge AI operates by processing data directly on devices, including phones, smartwatches, and security cameras. This reduces latency, eliminating the need for waiting for responses, and enables your technology to react instantly. For instance, if your voice assistant can turn on lights even when Wi-Fi is down, that's Edge AI at work. Since personal data stays on your device, it's far less vulnerable to hackers or leaks, giving you peace of mind in a world full of data breach news.

In everyday life, Edge AI pops up in ways you might not notice. Smart thermostats like Nest monitor your comings and goings, adjusting the climate based on your habits, eliminating the need for constant cloud communication. Over time, these helpers quietly save energy and money. Voice assistants process simple commands ("set a timer," "play jazz") on the device, keeping your requests private and delivering quicker responses so you're not left waiting or repeating yourself.

Wearables have also embraced Edge AI. Fitness trackers and smartwatches now analyze health metrics, such as heart rate, sleep cycles, and steps, directly on your wrist rather than on a remote server. This means faster feedback, instant alerts or trends, and greater privacy, as your sensitive data remains with you. In workplaces, Edge AI secures sensitive information: smart badges can monitor occupancy or air quality and flag issues immediately, all while keeping access logs under your control.

Security cameras with Edge AI instantly recognize faces and distinguish between strangers and family members. Footage stays local, so it's not floating online, lessening the risk of leaks or hacks. Your camera immediately alerts you if someone unfamiliar appears, rather than uploading every moment to the cloud.

Edge AI gives you better control of your info. Since raw data rarely leaves your gadget unless you allow it, you don't have to guess what companies might do with your recordings or video feeds. For parents, this means baby monitors that detect sleep patterns without streaming sensitive footage online. For everyone, tools like smart doorbells become less worrisome, knowing video isn't traveling to some private company's servers.

Edge AI is evolving quickly. One promising trend is federated learning, where devices learn together by only sharing minor, anonymized updates, never your raw data. Picture many fitness trackers collaborating to improve health predictions yet never exposing individual users' details. Healthcare wearables are beginning to use this method, helping spot arrhythmias while preserving patient privacy.

The rollout of 5G networks is further fueling the growth of Edge AI. With higher speeds and reduced lag, devices can coordinate instantly. Think of cars that swiftly process sensor data to avoid collisions or factories where machines can spot safety hazards and alert staff, all locally and immediately.

Miniaturization, the trend dubbed "tiny AI," is allowing powerful models to fit into chips the size of a postage stamp. This enables low-power devices, such as wildlife trackers or medical implants, to remain innovative and efficient, handling sensitive monitoring without a constant external connection. A tiny sensor could track endangered animals without interfering, or a medical implant could monitor health markers and notify doctors only when needed.

Edge AI isn't just hype or future talk. It's about brighter, smoother, more private digital experiences right now. Instead of devices leaving you in limbo while a distant server catches up, your technology responds instantly, with your data staying largely under your control. From daily routines to healthcare and home security, the future points to a world where devices truly work for you, privately and promptly.

How Edge AI Works

You ask your smart speaker to turn off the lights. With Cloud AI, your request travels to a remote data center and back, causing lag and exposing your data. With Edge AI, the request stays in your living room and is processed on the device for instant action and private data. That's the key difference between Cloud AI and Edge AI. Local intelligence is quicker and more secure than sending your data all over the world.

Affordable Tools You Can Deploy Today in AI for Small Businesses

Running a small business often feels like spinning plates, where you have sales calls, bookkeeping, marketing, and customer questions all at once. Suppose you've ever wished for help but are worried about the cost or complexity of technology. In that case, AI might finally bring some relief. You don't need a computer science degree or an IT team to start, either. The latest wave of AI tools is designed for everyday users who want real results quickly. Let's look at where intelligent automation can have the most significant impact: sales, customer service, operations, and marketing.

Start with sales. Customer relationships make or break a business, but keeping track of every call and lead is tough. AI-powered CRM platforms, such as *Zoho CRM* or *Salesforce Einstein*, do more than store contact information. They nudge you to follow up at just the right time, flag hot leads based on patterns from past deals, and even score which prospects are most likely to close. For a busy owner juggling everything, these nudges mean you won't lose a big sale just because you missed an email on a busy morning. Setup is straightforward: import your customer list, connect your email account, and let the AI start analyzing patterns. The dashboards are clear, and you can see which deals need attention today, not next week.

Operations can eat up hours with repetitive tasks, especially bookkeeping. *QuickBooks* now offers AI-driven categorization that automatically sorts expenses and income, learning your preferences as you use it. Instead of spending Friday nights sorting receipts or stressing over tax time, you review flagged transactions and approve suggested categories. This frees up time for high-value tasks, such as product development or client meetings. It also reduces costly errors that can slip in when you're tired or distracted.

Customer service is another area where AI can provide relief. Deploying a chatbot, such as *Tidio* or *Chatfuel*, on your website allows you to answer common questions 24/7. Visitors receive instant assistance finding store hours, product information or tracking their orders, which keeps them satisfied and boosts conversion rates. Setting up a chatbot doesn't require coding; most platforms offer drag-and-drop builders with templates for various industries, including retail, consulting, and services. You can even connect these bots directly to *Shopify* or *WooCommerce* with just a few clicks, automatically syncing inventory updates and order statuses.

Marketing might seem daunting if you're not a creative pro, but AI makes it easier to compete with bigger brands. *Mailchimp's* AI features recommend the best times to send emails based on when your audience usually opens them. The system predicts which subject lines will catch the most eyes. It automatically segments your list for targeted campaigns, eliminating guesswork and ensuring precision. *Buffer's* AI-powered insights eliminate the guesswork of social media scheduling by suggesting optimal posting times and analyzing the types of posts that drive the most engagement for businesses like yours.

Plug-and-play is the name of the game for these tools. Most offer free trials or affordable monthly plans that scale with your growth. For example, Tidio's free plan meets the needs of most startups; *Mailchimp* allows you to start with basic automation and upgrade as your list grows. The best-fit scenario? Select tools that address your current needs without overwhelming you with features you'll never use. If scheduling and follow-ups keep you up at night, prioritize CRM and email automation as your first step to improve your productivity. If it's managing customer questions at odd hours, a chatbot delivers instant impact.

Setting up these platforms takes less effort than you think. Let's say you want to connect a chatbot to your Shopify store: sign up for a Tidio account, follow the step-by-step integration guide (which typically involves copying a code snippet into your store's theme), and then customize the chatbot template to fit your brand's voice and FAQs. Test it by asking standard questions and tweaking responses as needed. To automate invoice generation from sales emails in *QuickBooks*, enable the AI categorization feature and link your business email account. Now, invoices are created automatically when an order is received.

Troubleshooting is part of the process, but it rarely requires more than a *Google* search or a quick chat with customer support. Common hiccups include syncing issues (usually fixed by reauthorizing permissions) or bots giving generic answers (solved by updating FAQs or adding new response paths). Most providers offer active online communities where you can find tips and solutions from other small business owners who have been there.

Real businesses everywhere are already seeing results with these tools, even on tight budgets. A local bakery in my city doubled its online orders within six months after installing an AI-driven marketing suite that sent reminders to loyal customers and automatically personalized birthday offers. They didn't hire new staff or spend hours learning new software; they followed simple onboarding steps and relied on templates provided by the platform. A freelance consultant I know saves hours each week using AI-powered contract review tools that flag risky clauses in proposals before sending them out to clients. Instead of combing through legalese line by line, she gets instant feedback on what needs attention and what's safe to approve.

It's worth noting that while AI can automate many tasks, it doesn't replace the personal touch, the handwritten thank-you note, or the creative spark behind a new campaign; these elements still matter. But by offloading repetitive tasks and providing intelligent recommendations when you need them most, these affordable tools enable you to focus on what only you can do: building relationships, serving customers, and growing your business in ways that feel authentic and sustainable.

Separating Hype from Real Value When Vetting New AI Products

If you've watched a flashy demo or browsed an AI company's landing page and felt unsure of what's real, you're not alone. Distinguishing handy AI tools from hype requires structure and sharp critical thinking. With new "AI-powered" apps appearing constantly, nearly anything can sound revolutionary. However, by watching for key patterns, you can separate genuine value from superficial buzz.

Start with a firm evaluation framework. First, look for proven use cases. Does the product solve real problems for businesses like yours or spout buzzwords? Seek transparency about how the tool works: does the vendor explain clearly what the AI does, what data it uses, and its decision process? A support team or an accessible knowledge base is a good sign that they're invested in their users. Use a checklist:

- Are there credible case studies or references?

- Is the privacy and security documentation clear?

- Can you reach real people for help?

If you can check these, you're on safer ground.

Common red flags accompany hype-heavy products. Be cautious of tools that claim to be "AI-powered" without explaining. If you can't figure out what it does or how it solves a problem, hesitate before investing time or money. Missing independent reviews or authentic customer testimonials is a warning; if all you see are paid articles or glowing "partner" quotes, dig further. A lack of clear privacy or security practices is a major red flag. If a vendor can't tell you where your data goes, how it's secured, or whether it trains its models, it's best to move on. Such gaps often indicate immature products or unserious companies.

When evaluating an AI tool for work, such as an HR analytics platform, never rely solely on marketing promises. Request a live demo using your data, not just a scenario. Ask challenging questions: "How does this system flag bias?" or "What happens if there's a mistake?" Listen for honest, straightforward answers, not empty sales talk. Afterward, ask for a pilot with a small group. Let real users test it; gather feedback on usability, accuracy, and whether it reduces workloads or adds busy work. See how responsive support is when issues arise. Prompt, helpful service always beats big promises.

Due diligence is ongoing. The AI market evolves rapidly; today's best-in-class product may become outdated in as little as six months. Set quarterly checkpoints to reassess your tools: do they remain a good fit, or are better options available? Maintain a team watchlist of promising new products to ensure decisions aren't based solely on one person, and your group remains vigilant to market shifts. Don't hesitate to retire tools that underperform; clinging to bad tech wastes time and opportunities.

Ongoing adaptation matters. AI products update and evolve rapidly. New features or changes can boost results or create new risks. Keep tabs on your vendor's roadmap by signing up for update emails or joining customer forums, if available. Teach your team to spot if tools misbehave or underdeliver. Rotate watchlist ownership to gain fresh perspectives. Different roles may flag issues others miss.

Connecting with outside voices is valuable. Join online communities or local user groups to learn from others' successes and pitfalls. Someone's bad experience with hidden fees or data leaks could help you dodge those issues, and a strong peer recommendation can give you confidence to try something new.

AI Product Evaluation Quick Reference

- Does the tool show proven use cases for businesses like yours?

- Are AI usage and data handling clearly documented?

- Can you find independent reviews or speak with customers?

- Is support responsive and easily accessible?

- Is privacy and security explained in plain language?

- Did you test the tool with your own data before scaling up?

- Are you reviewing the tool's fit quarterly to stay current?

This structured approach keeps you grounded when hype abounds. You'll waste less time, make better recommendations, and push back confidently when something doesn't add up.

As you continue to explore AI trends, remember that critical thinking and adaptability are more important than any single feature. The next chapter will focus on building practical skills for communicating about AI, whether you're advocating for adoption at work or explaining the difference between hype and genuine value to friends and colleagues.

Chapter Eight

Actionable AI Playbooks for Building, Buying, or Integrating AI Solutions

Should You Build or Buy? Making the Right AI Investment

Imagine you're in a meeting, coffee in hand, when someone asks, "Are we building our own AI or buying it off the shelf?" Suddenly, all eyes are on you. Maybe a competitor launched a custom chatbot, or your inbox is flooded with "plug-and-play" AI pitches. There's real pressure, plus the fear of missing out. However, there's no single correct answer here; it's more like choosing between a ready-to-drive car and a custom-built one.

First, examine your team's technical depth. Do you have in-house engineers with AI expertise, or is your IT already stretched to maintain business operations? If your team is strong, data scientists, developers, and project managers who love experiments, building your own solution may make sense. It offers flexibility, allowing you to tailor the tech to your workflow. But custom builds take significant time, money, and effort. Developing and testing can take months (or longer). Ongoing maintenance, updates, and new hires are part of the bargain. For a tiny startup, this can quickly tax resources.

Buying an existing AI product, on the other hand, enables you to get operational quickly. Off-the-shelf solutions are typically tested, supported, and often accompanied by onboarding services. Suppose you need quick results, such as avoiding customer loss due to slow service or competing with automated competitors. In that case, buying can help you close the gap quickly. The upfront costs are lower than building from scratch. But you'll sacrifice customization, risk data silos, and be tied to another company's roadmap. Highly regulated or niche processes might not be well-served by generic solutions.

Consider time-to-value too. For fast wins, automated scheduling, email triage, and buying make sense. If your aim is long-term market differentiation (such as a brand-new product or patented process), the building might be worthwhile. Compare long-term costs: subscriptions and licenses add up for store-bought tools, but custom builds need steady investment in updates and staff. Look at the big picture. Sticker prices don't always reflect future expenses.

Decision Matrix for Build vs Buy Scenarios

Let's review sample scenarios. A small startup with limited resources and no AI engineers almost always benefits from buying. For instance, an e-commerce store in need of shopper recommendations can license a proven AI engine for rapid results without heavy lifting. By contrast, a large enterprise with a sophisticated R&D team might build a custom AI tool to process massive datasets and roll out exclusive features.

Industry matters, too. Suppose a healthcare provider needs diagnostic AI that meets stringent privacy laws; purchasing a turnkey solution is a viable option. In that case, FDA-approved solutions can skip years of risky development. Meanwhile, a retailer with unique products might build in-house for precise control over recommendations.

Case Study

Take a mid-sized retailer that tried building its own recommendation engine; it misjudged the effort, ran into messy data issues, and fell far behind schedule with rising costs. In comparison, a healthcare provider who purchased an off-the-shelf diagnostic system halved deployment time and improved patient triage immediately.

However, pure build or buy isn't the rule. Many organizations blend both: purchasing foundational AI and customizing certain features as their needs and teams grow. This hybrid approach enables smaller teams to stay nimble and provides larger ones with flexibility.

When shopping for AI, prepare a thorough RFP (Request for Proposal) to clarify must-have features, integration points, pricing, support, and compliance requirements. Ask for detailed security information and customer references. If building, audit your team's skills, what do you have and what's missing? How big is the learning curve? Allocate a reserve budget and time for pilot projects; even small proofs of concept can reveal gaps early.

When collaborating with partners, either vendors or consultants, ask direct questions: How portable is our data? What happens if we switch providers? Is the code auditable? How often are models updated for fairness and accuracy? Early transparency means fewer surprises later.

Remember, you don't have to lock into one path. Many teams begin by purchasing and adding custom layers as they gain skills and confidence. The key is matching your approach to real business needs: speed versus customization, upfront cost versus strategic investment, and generic versus bespoke. Don't let buzzwords blind you. Focus on what actually brings value to your organization.

Checklist for Testing AI in Your Org while Running a Pilot Project

Piloting AI before broad deployment is like dipping your toes in the pool instead of diving in. It's about getting practical answers to "what if?" questions and uncovering potential issues early on. By starting small, you can minimize risk, prevent major disruptions, give your team a safe space to adapt, and discover how your own data performs with AI. Testing at a limited scale enables you to see how new workflows integrate into daily routines and gather honest feedback before anything is finalized.

Begin your pilot with a clearly defined and manageable use case. Don't try to tackle everything at once. Choose an area with clear pain points and measurable outcomes, such as customer support tickets or invoice processing. Define what success will look like:

- Do you want faster response times, fewer errors, or better customer ratings?

- Write your key performance indicators (KPIs) in plain language so everyone understands them.

Assemble your pilot team with a mix of technical experts and end users, those who know the workflow best. Diverse roles help spot issues early that management might miss. Assign clear responsibilities:

- Who runs the test?

- Who collects feedback?

- Who makes decisions?

Prepare your data carefully, eliminate duplicates, fill in any gaps, and ensure that privacy rules are applied. Insufficient data will sink a pilot before tech issues do. Determine which metrics to track and record the current baseline. For example, if your goal is to achieve faster invoice approvals, document the current approval times before the pilot so you can compare the results. Without this baseline, you can't measure improvement.

Two common reasons pilots fail are scope creep and unclear ownership. Avoid adding features or new departments during the test; stick to your original goals. Designate a single "pilot owner" to make decisions and maintain focus. Collect feedback from all user levels, not just management. Make it easy for frontline staff to say what works and what doesn't. Their feedback is key to determining whether you scale up or need to go back and adjust.

Baseline measurements are essential. If you automate customer replies but haven't tracked current average response times, you won't know if you've improved. Gather this data upfront. Hold regular check-ins, weekly or biweekly, so everyone can share updates and blockers.

When the pilot ends, make decisions based on clear evidence, not guesses. Use a simple review structure: restate the goal ("Reduce support ticket response time by 30%"), note what changed, show before/after numbers, and summarize user feedback. Document surprises: maybe AI flagged more fraud but also generated more false alarms. Conclude with a go/no-go call: Should the AI tool be expanded, or does it require further development? Capture what you learned so future pilots benefit.

A quick example from a financial chatbot pilot: The goal was to answer 60% of routine queries automatically within two minutes. Previously, only 25% of queries were handled by humans within five minutes. After six weeks, the bot resolved 72% of routine questions in under two minutes, with steady customer satisfaction. However, some users felt bot answers were too generic for complex queries, highlighting a need for clearer escalation to humans.

Pilot Review Checklist

- Clear project scope and KPIs

- A cross-functional team with defined roles

- Baseline data documented

- Data cleaned, privacy maintained

- Regular feedback sessions

- User input from all levels

- Before-and-after metrics tracked

- Surprises and blockers logged

- Final go/no-go recommendation

- Lessons learned captured

Piloting enables you to learn rapidly, build trust in new technology, and set the stage for smoother and broader rollouts later.

API Basics for Non-Techies for Integrating AI with Existing Software

"API integration" may sound intimidating, but it's really like using a universal adapter for electronics: APIs (Application Programming Interfaces) enable different software to communicate, much like connecting a European gadget to a US socket. When you connect an AI tool to Slack, your CRM, or email, it's the API making the handshake. You don't need to be technical to understand this. Just know which tools need to be connected and what happens once you do.

Most businesses rely on a combination of software, including *Slack* for chat, *Salesforce* for customer data management, and *Google Workspace* for document collaboration. Before integrating AI, inventory these tools and note which are cloud-based or mention "integrations" or "add-ons." That's your starting point. For example, if you want to perform real-time customer sentiment analysis, such as flagging negative messages in *Slack*, verify that both *Slack* and your sentiment analysis tool support APIs or built-in connectors (most modern systems do). This is like confirming all your gadgets fit existing power outlets or knowing which adapters you'll need.

Connecting the software is usually straightforward. On the AI tool dashboard, look for options such as "Integrations" or "Connect App." You'll typically be prompted to copy an API key (an extended code unique to your account) and paste it into the right spot in another service, such as *Slack* or your CRM settings. For more complex setups, like monitoring only specific *Slack* channels, you'll log in to both services, grant permissions, and specify preferences. The AI tool will then pull live data, analyze content, and send back results or alerts (like letting you know if negative sentiment spikes). If things break, platforms typically offer troubleshooting guides or wizards to help users resolve the issue.

No-code and low-code integration platforms, such as *Zapier* or *Make,* eliminate the need for programming. These services offer drag-and-drop interfaces that connect your apps, much like snapping together building blocks. Want new web form leads to be scored by AI, entered in your CRM, and pinged to Slack? You set each action as a "trigger," connect the relevant apps, and define what happens next. The platform handles syncing and background logic. In sales and marketing, CRM plug-ins often provide AI features, such as automatic lead scoring or smart reminders, that can be easily installed and enabled with no technical skills required.

Security is crucial. Every API connection is a potential entry point, so always secure your integrations. Use trusted API keys or OAuth (which authorizes access without sharing your actual password). Never send sensitive keys via email or chat. They're like giving out a spare house key. Restrict each tool's permissions to only what it really needs (avoid full admin access if possible). Examine your vendor's security documentation to look for details on how data is stored and encrypted, as well as who has access to it. If a vendor can't answer basic security questions, view that as a warning sign.

Quick Integration Readiness Checklist

- List your current software tools and confirm that they support APIs and integrations.

- Identify which workflows would benefit from AI features.

- Utilize no-code or low-code platforms if you prefer to avoid hands-on coding.

- Only copy API keys from trusted sources; store them securely.

- Grant the "least access" permissions required for integrations.

- Review the vendor's documentation covering data security and handling procedures.

- Test the setup with non-critical data before going live.

- Set up alerts in case an integration fails or unusual activity is detected.

- Document who set up each integration and when for easy team knowledge sharing.

APIs enable you to automate routine tasks and unlock AI capabilities without relying on IT constantly. As business tools become more modular, understanding how APIs work and how to manage integrations securely empowers you to control and customize workflows smoothly without getting caught up in technical complexity.

Managing Change to Get Your Team Onboard with AI Adoption

Introducing AI at work isn't just about installing new software; it's about how people respond to change. Projects can stall if teams aren't engaged, as adoption depends on their buy-in, not just the technology's capabilities. It's common for teams to be skeptical or anxious; some worry about job security, others resist learning new tools, and some distrust "black box" decisions. Dismissing these reactions will undermine your efforts.

Communication is crucial. Start by mapping out everyone affected, including frontline staff, managers, IT personnel, and executives. Tailor your message to each group. For the initial announcement, be open and practical: "We're introducing an AI assistant for customer inquiries to reduce repetitive work and free you up for more complex tasks. No jobs are being cut; this is to make your workload easier." Soon after, host a Q&A so employees can ask questions and share concerns. Tackle tough topics honestly, and explain "why" the change is happening, not just what's changing. When people understand the purpose and feel their input matters, they become more engaged.

Establish a genuine feedback loop, not just a suggestion box that goes unchecked. Create a dedicated *Slack* channel or email alias for rollout feedback, and make sure leaders respond to questions and act on good ideas. Communicate regularly: "Here's what we heard. Here's what we're changing." When team input is acknowledged and acted upon, resistance tends to lessen.

Training is critical. Don't rely solely on a manual. Offer live, hands-on workshops where participants can try out the AI tool in realistic settings and record these sessions for those who are unable to attend. Provide accessible digital training modules for self-paced learning. Encourage peer support by pairing early adopters with colleagues less comfortable with the new system. Host "AI office hours" each week, where anyone can raise questions or concerns, regardless of their complexity. Leadership involvement is key; when managers and peers adopt the tech enthusiastically, others follow.

It's essential to address skepticism and fear directly. Be transparent about your intentions and the implementation timeline, especially if people are concerned about job loss. Share pilot project examples where AI has simplified repetitive tasks without eliminating jobs; let those affected share their experiences. Emphasize that AI handles repetitive work, allowing staff to focus on problem-solving and building customer relationships. Address ethical concerns up front: explain your efforts to ensure fairness, avoid bias, and protect data. Promoting two-way communication is vital; don't hand down AI changes as top-down edicts.

Make sure employees don't feel devalued or like guinea pigs. Celebrate every milestone, no matter how small: "The chatbot is now answering 30% of tickets, saving the team from 200 repetitive emails this week!" Publicly recognize those who help others adapt and highlight successes in meetings or newsletters. Share stories from employees whose workflows improved or who found new opportunities with AI. This builds momentum and a culture where innovation feels normal, not threatening.

Monitor for burnout or "AI fatigue," especially if multiple new tools launch quickly. Stagger rollouts, check in with teams regularly and adjust the pace based on both feedback and metrics. Slowing down at times can speed up success, as adoption is more effective when everyone feels involved and supported.

Communication Plan Template

- **Stakeholder mapping:** Identify everyone affected by the rollout.

- **Initial announcement:** Use clear, honest language.

- **All-hands Q&A:** Provide space for real questions and concerns.

- **Feedback Loop:** Set up a dedicated feedback channel or email address.

- **Training plan:** Schedule live workshops, offer digital modules, and provide peer support.

- **Ongoing updates:** Regularly summarize feedback and communicate changes.

- **Recognition:** Celebrate wins and acknowledge those who help others adapt.

- **Regular check-ins:** Monitor for stress and adjust rollout speed as needed.

Measuring AI Impact through KPIs, ROI, and Success Stories

After rolling out your AI tool, whether it sorts invoices, answers chat messages or flags maintenance issues, the big question is: how do you know it's making a difference? Choosing the right metrics is crucial. Avoid focusing solely on technical statistics, such as "model accuracy" or "server uptime," unless they directly tie to your business goals. The most valuable metrics are those that reflect meaningful improvements, such as time saved, errors reduced, customer satisfaction, or increased sales. For example, track response times and satisfaction ratings for a customer service bot. For invoice automation, monitor average approval times and error rates to optimize efficiency. For marketing AI, look at conversion and click-through rates. These numbers are already part of your regular business assessments, so you don't need a tech background to interpret them.

Calculating ROI for an AI project is straightforward if you maintain a clear perspective. ROI compares the benefits to the costs: (Value created – Costs) / Costs. Imagine your accounts team spends 20 hours per week on invoices at $30 per hour. After implementing AI automation, the time drops to 5 hours, resulting in a savings of 15 hours or $450 per week. If the tool costs $400 a month, that's $1,800 saved versus $400 spent, giving an ROI of (1,800 – 400) / 400 = 3.5 or 350%. But don't focus only on hard savings. Consider the following soft benefits: fewer late payments, happier employees (thanks to less repetitive work), and improved customer experiences. These intangible gains, while harder to quantify, can be just as valuable as the dollars you save.

Complex data is powerful, especially when you can show clear improvements. For instance, a manufacturing plant that adopted AI for predictive maintenance saw unplanned machine downtime drop by 40%. The system identifies equipment at risk before breakdowns, reducing production delays, lowering overtime costs, and improving team morale by eliminating emergency repair scrambles. Or consider a marketing team using AI to segment email lists where open rates climbed from 18% to 28%, and sales conversions doubled within a quarter. These results tell compelling stories of real change, making it easier to win support for more AI initiatives.

Effective reporting is as important as the improvements themselves. Don't let your AI's impact get lost in endless spreadsheets. Visualize your results by creating a "before and after" dashboard that displays KPIs side by side, average response times, error rates, and customer ratings. Use simple color-coded graphs to spotlight progress at a glance. For leadership, provide a concise executive summary that covers what changed, the associated costs, and recommended next steps, using clear and concise language without jargon. When talking to your team, celebrate specific milestones ("Jess cut her ticket backlog by 80%!") and share lessons learned ("We missed the first goal but adjusted after retraining the model"). Use a consistent template with fields for key KPIs, anecdotes, and progress charts.

To get started, try making a simple dashboard using spreadsheet software or a free online tool. Set up three columns: Metric (like task completion time), Before AI (your baseline), and After AI (your latest numbers). Include a notes section for surprises, such as a drop in customer complaints or staff reporting lower stress levels. Keeping these reports consistent makes it easier to share results with both teams and leadership.

Ultimately, let your results prove the AI's value. The businesses that get the most from AI aren't always those with the fanciest tech but those who track what's meaningful and communicate it clearly. Improvements such as fewer late shifts, reduced crisis management, or increased sales from more intelligent targeting should be showcased and celebrated. Sharing these wins builds momentum for future projects and helps keep everyone invested in ongoing innovation.

Maintaining and Updating AI Systems to Avoid Model Drift

AI systems are not set-and-forget machines. They need regular attention, just like any piece of software or equipment you rely on every day. One of the biggest headaches in real-world AI is model drift. This occurs when your AI model begins to lose its accuracy or usefulness because the world around it has undergone changes. Your e-commerce recommendation engine was spot-on last season. Still, now, after a few months and the emergence of new trends, it starts suggesting products that simply aren't selling. Suddenly, what worked perfectly loses its edge, and users notice. That's model drift in action when your AI's predictions or suggestions no longer match the new data coming in.

Why does this matter so much? Your business keeps evolving: customers develop new tastes, competitors roll out fresh products, or external events shift priorities overnight. Suppose your AI is still making decisions based on stale patterns. In that case, you risk missed opportunities, lower engagement, or even costly mistakes. In finance, this could mean missing fraud signals; in healthcare, it might mean outdated diagnostics. It's not just about keeping up. It's about staying relevant and trustworthy.

So, how do you spot drift before it causes headaches? Begin by establishing a routine for monitoring performance. Set up regular sampling and evaluation, either monthly or quarterly, depending on the pace of your business. Compare your AI's current predictions to actual outcomes. If the accuracy drops or error rates creep up, that's your signal something's off. Some teams set automated alerts that flag sudden changes in key metrics, so you don't have to keep an eagle eye on dashboards 24/7. The right tools can highlight anomalies or shifts in data distribution that hint at drift long before it becomes visible to end users.

Retraining your model is the cure for drift. This means feeding your AI fresh data and letting it learn from new patterns. Schedule these updates on a routine basis; consider refreshing every quarter for stable workflows or every month when things move quickly (such as in retail or with trending apps). Always keep a version history for your models. Reasonable version control allows you to track changes and roll back to a previous setup if something goes wrong after an update. Never overwrite your old model and hope for the best; instead, keep backups and test new versions on a sample of real-world data before rolling them out fully.

Updating AI safely requires a step-by-step approach. First, collect recent data reflecting new business realities. Clean and label it correctly. Garbage in still means garbage out. Train a new model version alongside the current one, checking both against recent results. If the updated version performs better, swap it in; if not, investigate what went wrong instead of pushing forward blindly. Always document the changes so future team members can follow the logic.

It's not just a tech team job. Maintaining AI health is a collaborative effort. Business leads bring context; what's changed in customer needs or regulations? Data scientists spot technical signals and recommend updates. Regular cross-functional review meetings keep everyone in sync. These don't have to be formal or stuffy. Monthly "AI tune-up" sessions work well: bring coffee, compare results, discuss any weirdness, and decide if retraining is due. Shared dashboards make it easy for everyone, not just data analysts, to identify if something appears to be off. When business and tech teams share ownership, issues are caught earlier, and everyone feels invested in the outcome.

To make drift management part of your culture, establish simple habits: schedule regular check-ins on your calendar, automate as much monitoring as possible, and build quick feedback loops for staff to flag any unusual outputs. Don't wait for significant failures; minor tweaks along the way save time and reputation.

Wrapping up this chapter, the big picture is clear: effective AI isn't just about flashy deployment, it's about thoughtful upkeep. Spotting and fixing model drift keeps your AI sharp and your business moving with confidence. The next chapter will look outward, exploring how to build an ongoing learning community around AI so you're always ready for what comes next.

Chapter Nine

Your AI Journey Using Community, Resources, and Lifelong Learning

Building Your AI Advisory Circle Consisting of Mentors, Peers, and Online Groups

Have you ever felt stuck, unsure if you're on the right track, and wished for someone to guide you? That moment, lost in tutorials and explainers yet still uncertain, highlights why a support network is essential. AI evolves quickly, and the fastest growth comes not from learning alone but from connecting with others on similar journeys. A strong advisory circle, including mentors, peers, and online groups, not only answers your questions but also builds your confidence, boosts motivation, and opens up new opportunities. Research confirms this: approximately 80%

of organizations now utilize AI, and those engaged in active communities learn faster and avoid common pitfalls.

Mentors are more varied than most realize. A technical mentor helps with neural nets or debugging. An industry mentor explains how AI fits into your field, be it retail, healthcare, or another area. A career mentor helps you identify opportunities, make informed transitions, and prevent burnout. Combining these perspectives is powerful. You might meet a technical mentor on *GitHub* or *Slack*, an industry mentor at a *Meetup* or webinar, and a career mentor via *LinkedIn* or even a previous boss. Aim for diversity: one viewpoint sets you on a path, but several show you your options.

How do you reach out? People often overthink the first message. Keep it concise, specific, and genuine:

Hi [Name], I'm inspired by your work in [AI topic/industry]. I'm beginning to learn about [specific area], and I value any advice for someone like me. Would you be open to a short virtual coffee chat this month? Thank you!

This message works, it's personal, researched, and asks for something concrete, not some vague mentorship. If they reply, book a short (15-minute) call and come with thoughtful questions. Afterward, send a thank-you and mention what you're trying based on their advice. To nurture the connection, schedule recurring "coffee chats" on a monthly or quarterly basis and use calendar reminders to prevent relationships from fading.

Connecting with peer groups and active communities is essential when you hit learning plateaus or need fresh perspectives. *Slack* groups, such as *the Data Science Society* and *AI Guild*, are full of people exchanging tips, sharing job leads, and offering honest feedback. On *Reddit*, the r/MachineLearning and r/ArtificialIntelligence communities blend news with lively discussion and real-world problem-solving. *Meetup.com* helps connect you with face-to-face events in your city, as well as those featuring "AI" or "machine learning." Even if you're not in a tech hub, virtual meetups unite people from all over the world.

Showing up online or at a meetup is just the beginning. The actual benefits come from active participation. Ask clear, specific questions (like, "Has anyone automated meeting notes with *Otter.ai* at work? What was your experience?"). Share your challenges or wins; others will respond with advice or share their own stories. Offer help, even if that means just sharing a great resource. Rotating "peer learning" sessions, where someone presents a short case study or tool each week, sparks lively discussion and shared advancement. Posting your case study or article for group feedback can clarify your thinking and bring insights from diverse perspectives.

Advisory Circle

List three mentors or peers you want to connect with this month, noting why each is compelling, and draft your first outreach message for each one. Then, pick two groups (*e.g., Slack, Reddit, Meetup*) to join this week. Pledge to ask one question or share a resource in each within seven days. This is more than networking; it's about creating a real circle of support for your AI learning. As you accomplish these tasks, you'll feel a sense of fulfillment and satisfaction, knowing that you're actively building your support network and enhancing your AI journey.

The people you surround yourself with shape the pace and depth of your journey. With the right mentors, peers, and community ties, every challenge gets easier and every achievement more meaningful. Choosing to step beyond solo learning might be your most brilliant move in the world of AI.

Use Curated Resource Lists to Stay Relevant Without Overload

Keeping up with AI news, tools, and learning materials can become overwhelming quickly. It's tempting to subscribe to every newsletter and save every 'Top 100 AI Tools' link, but this often leads to inbox overload and decision fatigue. Instead, you need a simple, repeatable system for filtering what genuinely matters, a structure that brings clarity, not chaos. The easiest way I've found is to divide resources into two buckets: **core** and **explore**. Having this structured system in place will bring you a sense of relief, knowing that you're not drowning in a sea of information but are focused on what truly matters for your AI journey.

Your "core" list is your foundational set of trusted sources that deliver consistent, relevant, and high-quality information. These include a few carefully selected courses, a key newsletter, and a handful of trusted blogs. For beginners, Andrew Ng's "AI for Everyone" breaks down complex concepts clearly and understandably. If you're focused on the business side, Harvard Business Review's AI pieces strike a balance between strategic insights and practicality. For those wanting technical depth, "The Batch" newsletter from DeepLearning.AI is up-to-date and accessible.

On the other hand, your "explore" list is for experimenting with new tools, listening to trending podcasts, or discovering interesting GitHub repositories you've heard about. Think of this as your sandbox for curiosity without distracting from your main learning path. I track both lists in Notion, but a simple Google Doc will do. Alongside each item, I leave a brief note: why it's on the list (whether it's a single-use resource or something ongoing) and what I hope to achieve from it. Every month or so, I review and trim anything that's lost value or become outdated.

The key is to keep your resource lists dynamic, not static. Regularly (every few months), revisit your core list. Ask yourself: Am I still learning here? Is this resource still current? Has something better appeared? This habit keeps your toolkit sharp and reduces "resource guilt" from unread articles or saved tools you'll never use. It also helps you notice shifts in your interests, such as moving toward more technical or example-driven content.

Evaluating new resources before you invest significant time is crucial. Start with freshness: In AI, even two-year-old material may be obsolete. Next, consider who created it. Is the author reputable? Are there positive reviews or commentary? Advanced readers should favor peer-reviewed sources (such as *arXiv* or *Springer*) over random blog posts for their technical depth. For courses, check user ratings on *Coursera* or *Udemy*. Older *GitHub* repos with many stars, and recent updates are more trustworthy. Still on the fence? Ask peers: "Has anyone tried this? Is it worth it?" You'll receive quick and honest feedback.

Here's a sample resource map by role and learning style to get you started:

- **Beginner:** "AI for Everyone" (*Coursera*), Kaggle's "Intro to Machine Learning" mini-courses, *YouTube* explainers of "AI: A Modern Approach."

- **Business-focused:** HBR's AI articles, "Practical AI" podcast, MIT Sloan Management Review's AI coverage

- **Technical:** "The Batch" (*DeepLearning.AI*), *Fast.ai* courses, the *Deep Learning Book* by Ian Goodfellow (for advanced study)

- **Visual learners:** 3Blue1Brown's neural network videos, infographics from *Towards Data Science*

- **Community-driven:** Curated *Reddit*/wiki lists, open-source tutorials with active discussions

To stay organized, set up a *Notion* or *Google Doc* log with columns for the source name, type (core or explore), update frequency, your rating, and notes on usefulness. This system makes it easy to identify what's working and where your learning needs to be adjusted.

Filtering resources isn't about limiting yourself; it's about focusing on what truly drives your growth. With a curated and regularly updated resource list, staying on top of AI becomes manageable and purposeful, not a race to keep up with every trend. You'll advance efficiently without getting overwhelmed by unread tabs or newsletters.

Participate in the AI Conversation at Conferences, Forums, and Webinars

Entering the bustling world of AI events can feel daunting at first, like walking into a gathering where you're unsure about the topics or faces. Yet, these events (whether in-person conferences or virtual webinars) are unmatched opportunities for learning and connection. They help you spot emerging trends, meet others wrestling with similar challenges, and discover creative approaches rarely found in articles or courses. By engaging directly, you tap into honest conversations, gain behind-the-scenes insights, and experience the excitement of collective curiosity.

There's a unique energy at in-person conferences: the buzz of ideas, impromptu networking over coffee, and serendipitous hallway chats that often spark new projects. Events like "Web Summit" and "NeurIPS" attract a mix of experts, researchers, and enthusiasts. At the same time, the AI Summit emphasizes practical, industry-focused applications. If travel isn't feasible, virtual events offer plenty of perks: you can participate from anywhere, avoid travel hassles, watch session replays, and interact with speakers in live chats or Q&As. Most events archive talks, allowing you to review key moments on your schedule.

AI events cater to a wide range of interests. For research-focused minds, NeurIPS offers a wealth of new concepts. If you're drawn to the intersection of business and technology, events like Web Summit and The AI Summit are full of actionable ideas. Those interested in sector-specific issues can explore summits such as the AI in Healthcare or AI for Good Global Summit, where experts tackle real-world challenges, including healthcare innovation, ethics, and social impact. On a local level, small meetups and hackathons provide close-knit opportunities to learn, collaborate, and build relationships, often right in your city.

To get the most from these gatherings, it pays to be strategic. Before an event, clarify what you hope to learn and who you'd like to meet. Your goals might be along the lines of "find three practical AI case studies for retail" or "connect with someone who's launched a chatbot." Update your *LinkedIn* profile and bring business cards. During the event, take concise notes; focus on key takeaways, memorable ideas, or actionable items rather than trying to capture everything. Photos of slides can serve as helpful visual reminders, provided they are allowed. After the event, promptly follow up with new connections and send short, specific messages referencing your conversation to help them recall you.

Greater value is achieved when you shift from being an attendee to a participant. Start by volunteering to help with logistics or moderating discussions, which will put you in deeper conversations and behind-the-scenes action. If you're ready, submit a proposal for a lightning talk or join a panel; most events welcome fresh voices, and sharing your story, even with its failures, can offer invaluable lessons to others. Supporting local meetups or helping to organize events is another way to deepen relationships and gain a deeper understanding of the AI community from an insider's perspective.

Live forums and webinars also offer rich opportunities for learning and networking, especially in asynchronous formats. Participate in webinar chats with insightful questions ("How did you deal with X challenge using limited data?"). Continue conversations in event forums, share your thoughts, or debate key points. Don't hesitate to direct-message individuals whose commentary or work interests you; these exchanges can evolve into collaborations or job prospects.

Pre-Event Goal Setting Worksheet

- Two topics or trends I want to learn more about:

- Three speakers or companies I'd like to connect with:

- One question I want to be answered before I leave:

- My preferred follow-up method (*LinkedIn*, email, *Slack*):

Use this worksheet beforehand to stay focused on your objectives rather than getting distracted by expo giveaways.

Participating in AI events as a listener, contributor, questioner, or volunteer helps you shape your learning journey while expanding your network and keeping your skill set fresh and relevant.

Keeping Up with AI News Using Trusted Sources and Newsletters

Trying to stay updated on AI can feel like trying to drink from a firehose. Every day, there's a new headline, another "breakthrough," a shiny tool launch, or a big-name warning about robots taking over. If you've ever gotten lost in endless tabs, newsletters piling up, and podcasts you meant to catch up on last month, you're not alone. News overload is a real phenomenon, and it can turn curiosity into stress. The trick is to flip the script: you want the news to serve you, not drown you. Instead of reading every trending article, start by selecting a few trusted sources that align with your interests and learning style. Then, set intentional boundaries. I give myself twenty minutes each morning for "AI news, "no doom-scrolling, no clicking every link. It's enough to catch significant shifts without losing half my day. Setting a timer helps me stay focused and keeps things under control.

Curating your news diet begins with selecting outlets that strive for balance and provide actionable insights. I recommend newsletters like "Import AI" by Jack Clark. He breaks down complex trends into plain language, providing just enough detail to keep you sharp without delving too deeply into the weeds. "AI Weekly" gives a straightforward roundup of what's happening, cutting through the hype. For audio, the "Lex Fridman Podcast" offers thoughtful interviews with AI leaders and skeptics alike, perfect for commutes or lunch breaks. "The TWIML AI Podcast" (This Week in Machine Learning & AI) digs into real-world cases and practical lessons, keeping things grounded and relevant. These sources rarely push a single agenda, which is refreshing in a space full of hot takes.

Now, organizing all this information so it doesn't become another mess is key. Automating your intake helps keep everything tidy. RSS feeds are your friend here; *Feedly* is a great starting point. Set up custom feeds for topics like "machine learning" or "AI ethics," and it pulls in only the content you request. Instead of dozens of emails hitting your inbox, newsletters get bundled; some apps even summarize them for you so you can skim the highlights fast. *Google Alerts* is another underrated tool. Set up a few for your top interests ("AI hiring," "explainable AI") and let the headlines come to you on your schedule, not the internet's. I use folders and tags to sort articles by priority or project, so nothing slips through, but nothing gets overwhelming.

Reading critically makes all the difference. Not every headline deserves your trust or your panic. When you spot something wild ("AI predicts earthquakes!"). Pause and check: Who's reporting this? Is there an actual study or just a press release? I like to jot down questions in my notes or flag stories that seem too good (or bad) to be true. Sharing these "let's discuss" stories in team *Slack* channels or group chats sparks good conversation. Sometimes, someone with more experience can spot what's hype and what's real in seconds. Creating your roundup document (a simple list of top stories, big questions, and takeaways) helps organize your thoughts, track patterns over time, and gives you something to refer back to later.

If you're feeling swamped by updates, remember that missing a few headlines won't set you back. The goal isn't to be first; it's to be thoughtful and informed enough to spot what matters for your work or personal growth. You'll find that by setting limits, choosing reliable sources, automating tasks where possible, and always reading with a healthy dose of skepticism, you can keep your curiosity alive without burning out. The proper news habits will help you spot signals among the noise and stay sharp as AI keeps evolving at breakneck speed.

How to Share Your Own AI Stories and Wins to Contribute Back

There's significant value in sharing what you've learned, even your missteps. You don't have to be an influencer or a recognized leader; anyone who's improved a process, solved an annoyance, or learned from a mistake has a story worth telling. Sharing experiences not only helps you build credibility with colleagues and potential employers, but it also opens doors for new opportunities and meaningful connections. People remember real stories far more than lists of skills and credentials. Saying, "I automated my scheduling and gained back an hour a day," is memorable; these narratives invite conversations, collaborations, and even future job offers.

Your contributions can take many forms: Write a blog post like "How I Used AI to Clean Up My Inbox" or "What I Learned from Trying Five Chatbots." LinkedIn articles can engage your professional network, and short posts with a simple screenshot and before/after summary often resonate even more. You might also consider running a live demonstration for a local group, hosting a virtual "show and tell" at work, or presenting internally to your team. An easy way to share "how I automated our monthly reports," for example, can quickly establish you as a go-to resource. If public speaking makes you nervous, small sessions with colleagues or friends are a good place to build confidence and share knowledge.

Good storytelling is about structure and honesty, not polish. Start by outlining the problem or challenge you faced. Then, walk through what you tried, including what worked, what failed, and any surprises along the way. A simple before/after format is effective: "Before, I spent an hour each week copying data; after my AI script, it takes just five minutes." Don't skip over what didn't go as planned; those details contain the most valuable lessons. Sharing "what I wish I'd known" or your biggest mistake encourages others to try, fail, and learn, too. Authentic, specific accounts resonate far more than generic success stories.

Supporting others is another powerful way to contribute. Volunteering as a mentor, even informally, deepens your understanding of the subject. You don't need to be an expert: having completed a course or set up a tool already puts you ahead of total beginners. Join a mentorship program in your field, or offer a session for newcomers in your local group or online community. Starting a "newbie-friendly" FAQ or resource page in a chat channel or forum can have a significant impact; sometimes, simply pinning a short list of tips is enough to help many beginners feel welcome.

Sharing both your successes and failures makes you part of a broader conversation. You help others avoid your mistakes and, in turn, raise your profile. People will start reaching out for advice or new opportunities; these connections can lead to job offers, partnerships, lasting friendships, and unexpected invitations to speak. To start, keep it small: post a brief "How I Automated X in my workflow" update on *LinkedIn*, or do a five-minute demo for your team about the tool that saved you hours. Regular, honest sharing builds lasting momentum. Over time, your stories inspire others, demystify AI in practical terms, and foster a culture where ongoing learning is valued and accessible.

If you're hesitant or unsure, remember that all experts started as beginners, often with awkward first posts or demos. Simply caring enough to share, even imperfectly, is what sets contributors apart from passive observers. Begin with one small, concrete story this month. The impact and the community that forms around you may surprise you.

The Living Blueprint for Updating Your AI Skills as the Field Evolves

AI is constantly changing, and so should your approach to learning it. Treating AI as a one-time study is a trap. What's popular now may be outdated soon. The most successful people in the field expect change and remain curious. To stay sharp and confident, commit to ongoing learning and professional development. Think of your growth as a living blueprint, constantly updated with each new challenge or shift in interest. Personal updates aren't just for your apps; they're for you, too.

Make ongoing skill updates practical by building a routine. Schedule a personal review every few months, whether you call it a quarterly audit or conduct it twice a year. Take a moment to assess where you are, what has changed, and where you want to go. I like using a simple

SWOT analysis (Strengths, Weaknesses, Opportunities, Threats) focused on AI skills. Identify your current strengths and weaknesses, exciting or functional new areas, and any gaps you've noticed in work or projects. The goal isn't self-criticism; it's to find growth opportunities and set one or two "next skill" targets for the coming months. You should get better at conversational AI, understand explainable models, or improve at integrating automation tools into your workflow.

Feedback is crucial. After completing an AI project, whether big or small, request honest feedback from colleagues, clients, or end-users. What worked well? What could have been easier? Valuable insights often come from those who have used your solution, not just observers. Don't just seek praise; look for what slowed people down or caused confusion. Combine this feedback with data: Did your bot reduce response times? Did your automation cut errors or save significant time? Track and measure tangible outcomes. Let these results guide the skills you focus on next. The aim is to build practical, impactful abilities, not just collect certificates.

Adaptability and curiosity aren't optional in AI; they're essential. When you hear about a promising tool or platform, check it out and sign up for early access or beta testing when available. Experimenting with new technology helps you distinguish real value from hype. Participating in hackathons or AI sprints is another way to push yourself: even if you feel underqualified, the hands-on learning and exposure to new ways of thinking are invaluable. If public events aren't appealing, follow a few insightful voices on social media, those who break down trends, spot fluff, and share practical insights. Watch for patterns, new skills, and inspiration for your next step.

This approach keeps learning manageable and engaging rather than overwhelming. You don't need to know everything. Just keep your living blueprint up to date with what matters to you right now. AI will continue to evolve, and so can you.

Quick Self-Assessment

Take ten minutes and jot down:

- One AI concept or skill you've learned in the past six months.

- One skill you're curious about or feel rusty on.

- One project or tool you'd like to try before the year's out.

- A piece of feedback (positive or negative) that made you reflect.

Choose one "next skill" to focus on for the next three months. Set calendar reminders to revisit your progress.

Remember: real progress comes from small, repeated steps and regular self-checks. Your blueprint is never final. It evolves with each attempt and a new lesson. Stay flexible and open-minded. In the next chapter, we'll explore how to apply your skills to achieve tangible results in your work and life, so get ready to transition from learning to doing.

Conclusion

So here we are, at the finish line, together. When you picked up this book, you may have felt curious, a bit cautious, or flat-out overwhelmed by all the AI talk swirling around you. If you've made it this far, let me say you've already done something that most people only talk about. You invested in yourself, your career, and your future. That's huge.

From the very first page, my promise to you has been simple: AI shouldn't be a locked box only a few can open. It shouldn't be a buzzword that makes you feel like you're missing out or falling behind. I wrote this book for adults like you, who are busy, goal-oriented, and often juggling work, life, and the endless stream of "new tech" headlines. My mission has always been to make AI practical, approachable, and, most importantly, empowering for you. No intimidation. No gatekeeping. These are just real wins you can grab today.

Let's take a moment to look back at the ground we've covered together. We began by breaking down what AI is (and isn't), shedding the sci-fi myths, and ditching the jargon, making room for genuine understanding. We explored industry case studies, ranging from healthcare, where AI is used for early disease detection, to marketing, where AI aids in personalized advertising, and to retail, where AI powers recommendation systems, so you can see AI at work in places that impact your life. We rolled up our sleeves and tackled hands-on projects, delivering immediate, shareable wins, such as smarter meetings, decluttered inboxes, automated workflows, and enhanced customer support.

We discussed upskilling and mapping out your learning journey, whether you're just starting or ready to pivot your career. We tackled ethics not as an afterthought but as a thread running through every example and exercise. You learned how to spot bias, build fairness into your AI projects, and prioritize privacy. We didn't shy away from the tough stuff, either, such as communicating AI, debunking workplace myths, and making complex ideas visual and accessible for everyone on your team.

We examined the trends (without the hype), learned how to identify value in new tools, and created actionable playbooks for integrating AI solutions into your daily workflow. We wrapped it all up by building a lifelong blueprint and community because AI isn't a one-and-done skill; it's a journey best taken together.

Here's what I hope sticks with you most: AI is for everyone. You don't need a computer science degree. You don't need to code. You don't need hours of free time or a fancy job title. What you do need is curiosity, a willingness to try, and a mindset that's open to learning. Visual guides, step-by-step checklists, and real examples are not just nice-to-haves; they're the heart of effective learning. And using AI responsibly isn't optional; it's the only way forward. The best results come from combining your judgment, values, and creativity with the power of innovative tools.

I want to celebrate your journey from confusion or even skepticism to a place where you can confidently say, "I understand this. I can do this. I can use AI to make my job easier, my day smoother, and maybe even my life a little better." This is not just progress; this is a significant transformation. You're no longer on the outside looking in. You've gained practical skills, confidence, and a clear game plan for what's next.

So, what's next? Here's my challenge: choose one win from this book and put it into action this week. Whether it's setting up an AI-powered note-taker for your next meeting, automating a repetitive task, or sharing your newfound knowledge with a colleague, don't keep your success to yourself. Please share it with your team, your friends, or our online community. The more you share, the more you'll learn, and the more you'll inspire others to join you on this journey.

And speaking of community, this is not a farewell. The online resource and case study library are vibrant, interactive spaces. I want to hear your stories, your questions, your frustrations, and your wins. Your feedback will help make this resource better for everyone who comes after you. So, don't be a stranger. Drop in, contribute, and help shape the next chapter of this ever-evolving journey. You're not alone in this.

Remember: this isn't just another technical manual or a book full of empty promises. This is your visual-first, ethics-driven, no-excuses-needed, practical AI playbook. It's built for your life, your work, and your goals. And it grows with you. You don't have to do this alone, and you don't have to know everything before you start.

As you move forward, know this: you're not just keeping up with the future. You're helping to shape it. The world needs more people who use AI thoughtfully, ethically, and creatively - people who ask the right questions, demand fairness and look for real impact over hype. I believe you're one of those people now.

So keep learning. Keep questioning. Keep sharing. Remember, the field of AI is constantly evolving, and staying updated is key to thriving in this domain. You have what it takes to thrive and to help others do the same in a world where AI is part of daily life. Thanks for letting me be a part of your journey. Now, go out there and show what practical, empowering AI can really do.

References

- Aryan. (2023, August 31). *8 Best AI Email Management Tools in 2025*. Hiver. https://hiverhq.com/blog/ai-email-management-tools

- Boesch, G. (2025, April 4). *Explainable AI (XAI): The complete Guide (2025)*. viso.ai. https://viso.ai/deep-learning/explainable-ai/

- Kanakia, A., Sale, M., Zhao, L., & Zhou, Z. (2025). AI in Action: Redefining drug discovery and development. *Clinical and Translational Science, 18*(2). https://doi.org/10.1111/cts.70149

- Marr, B. (2023, July 5). *Debunking AI myths: The Truth behind 5 Common Misconceptions*. Forbes. https://www.forbes.com/sites/bernardmarr/2023/07/05/debunking-ai-myths-the-truth-behind-5-common-misconceptions/

- Rebelo, M. (2025, March 5). *The 8 best AI meeting assistants in 2025*. Zapier. https://zapier.com/blog/best-ai-meeting-assistant/

- Reche, A. (2024, December 26). *16 best AI tools to increase your productivity 2024*. VoIPstudio. https://voipstudio.com/blog/ai-productivity-tools/

- Simba 7 Media. (2024, July 28). *AI and GDPR Compliance in Marketing Practices: Balancing Innovation with Data Protection*. https://simba7media.com/ai-and-gdpr-compliance-in-marketing

-practices-balancing-innovation-with-data-protection/

- *What is the difference between AI vs. machine learning vs. automation?* | *edX.*   (n.d.).   edX. https://www.edx.org/resources/what-is-the-difference-between-ai-vs-machine-learning-vs-automation

- Zhou, V. (2019, March 3). Machine Learning for Beginners: An Introduction to Neural Networks. *victorzhou.com.* https://victorzhou.com/blog/intro-to-neural-networks/

- Awan, A. A. (n.d.). *7 AI portfolio projects to boost the resume - KDNuggets.*   KDnuggets. https://www.kdnuggets.com/7-ai-portfolio-projects-to-boost-the-resume

- Birkins, J. (2024, November 25). 16 AI workflow Automation platforms for No-Code AI workflows. *Medium.* https://medium.com/@joycebirkins/16-ai-workflow-automation-platforms-for-no-code-ai-workflows-2118eb57069f

- *Fairness: Equality of opportunity.* (n.d.). Google for Developers. https://developers.google.com/machine-learning/crash-course/fairness/equality-of-opportunity

- *How can you explain AI to stakeholders without technical knowledge?* (2024, July 22). www.linkedin.com. https://www.linkedin.com/advice/0/how-can-you-explain-ai-stakeholders-without-technical-ac1jf#:~:text=Explaining%20AI%20to%20non%2Dtechnical,can%20help%20demystify%20its%20functionality

- *Human-in-the-loop AI: 4 best practices for workflow automation | Tines.* (n.d.). https://www.tines.com/blog/humans-in-the-loop-of-ai/

- Jungco, K. G. (2025, May 29). *9 Best AI Certification Courses to*

Future-Proof your Career. Datamation. https://www.datamation.com/artificial-intelligence/artificial-int elligence-certifications/

- Kekare, D. (2025, May 25). How to Transition into an AI Career from a Non-Tech Background: 7 Practical Steps. *Data Expertise.* https://www.dataexpertise.in/ai-career-transition-non-tech-back ground/

- Marr, B. (2023, July 5). *Debunking AI myths: The Truth behind 5 Common Misconceptions.* Forbes. https://www.forbes.com/sites/bernardmarr/2023/07/05/debun king-ai-myths-the-truth-behind-5-common-misconceptions/

- Martin, G., & Martin, G. (2024, November 23). *The role of AI in Enhancing Visual Storytelling: Revolutionizing content creation.* PRO EDU. https://proedu.com/blogs/photoshop-skills/the-role-of-ai-in-en hancing-visual-storytelling-revolutionizing-content-creation#:~: text=In%20visual%20storytelling%2C%20AI%20algorithms,mo re%20interactive%20and%20personalized%20narratives

- Ramki, H. (2024, November 26). *11 actually great elevator pitch examples and how to make yours.* https://zapier.com/blog/elevator-pitch-example/

- Salon, D. S. (n.d.). *Top 13 Data Science and Machine Learning Slack communities.* https://roundtable.datascience.salon/top-data-science-machine-learning-slack-communities

- Technologies, D. (2022, September 20). *Real-life examples of discriminating Artificial intelligence - Datatron.* Datatron. https://datatron.com/real-life-examples-of-discriminating-artific ial-intelligence/

- *11 AI communities that will accelerate your learning in 2025 | DigitalOcean.* (n.d.). https://www.digitalocean.com/resources/articles/ai-communities-for-learning

- Admin. (2024, January 5). AI and ML Development: Leverage the Power of AI & ML for Your Business with Bitdeal's Development Services. *Bitdeal.* https://www.bitdeal.net/ai-ml-development

- Glover, E. (2024, December 3). *26 Top Generative AI tools.* Built In. https://builtin.com/artificial-intelligence/generative-ai-tools

- Gulfraz, A. (2024, July 8). *Top AI conferences to attend in 2024.* https://www.linkedin.com/pulse/top-ai-conferences-attend-2024-ayesha-gulfraz-druvf

- Jake. (2025, April 10). *Best AI newsletters and podcasts.* Delighted Robot. https://delightedrobot.com/2025/04/10/best-ai-newsletters-and-podcasts/

- Kanerika, & Kanerika. (2025, May 8). *How to launch a successful AI pilot Project: A comprehensive guide.* Kanerika. https://kanerika.com/blogs/ai-pilot/

- Lewis, G. (2024, February 23). *How four small businesses are getting a bang for their AI buck.* Raconteur. https://www.raconteur.net/technology/four-ai-case-studies

- *Library: Norwalk AI in Education: Ethical Considerations & Copyright.* (n.d.). https://library.ctstate.edu/c.php?g=1412708&p=10476014#:~:text=Misinformation%20and%20Deepfakes%3A,trust%20and%20truth%20in%20media

- nopAdvance. (n.d.). *Understanding Gen AI: the next frontier in artificial intelligence. Send Large Files | Transfer Infinite | hub of bulk file sending.* https://transferinfinite.com/info/understanding-gen-ai-the-next-frontier-in-artificial-intelligence

- Quagliotti, F. (2024, May 15). *Navigating the AI implementation journey: Buy or Build?* Tryolabs. https://tryolabs.com/blog/buy-vs-build-ai-a-guide-for-decision-makers

- Sand Technologies. (2025, May 30). *From Investment to Impact: A Practical Guide to Measuring AI ROI.* https://www.sandtech.com/insight/a-practical-guide-to-measuring-ai-roi/

- *What are the best ways to find a mentor in Artificial Intelligence (AI)?* (2023, December 4). https://www.linkedin.com/advice/3/what-best-ways-find-mentor-artificial-intelligence-cxlee

- White, T. (2024, April 22). API Integration: A Non-Technical Guide for Beginners. *Neighbourhood.* https://www.nbh.co/learn/api-integration-a-non-technical-for-beginners

- XenonStack. (n.d.). *Edge AI for smart home applications.* https://www.xenonstack.com/use-cases/edge-ai-for-home-applications

www.ingramcontent.com/pod-product-compliance
Lightning Source LLC
Chambersburg PA
CBHW071420210326
41597CB00020B/3593